Structural Health Monitoring for Advanced Composite Structures

Computational and Experimental Methods in Structures

ISSN: 2044-9283

Series Editor: Ferri M. H. Aliabadi *(Imperial College London, UK)*

This series will include books on state-of-the-art developments in computational and experimental methods in structures, and as such it will comprise several volumes covering the latest developments. Each volume will consist of single-authored work or several chapters written by the leading researchers in the field. The aim will be to provide the fundamental concepts of experimental and computational methods as well as their relevance to real world problems.

The scope of the series covers the entire spectrum of structures in engineering. As such it will cover both classical topics in mechanics, as well as emerging scientific and engineering disciplines, such as: smart structures, nanoscience and nanotechnology; NEMS and MEMS; micro- and nano-device modelling; functional and smart material systems.

Published:

More information on this series can also be found at http://www.worldscientific.com/series/cems

(Continued at end of book)

Computational and Experimental Methods in Structures – Vol. 8

Structural Health Monitoring for Advanced Composite Structures

Editors

M H Ferri Aliabadi
Z Sharif Khodaei

Imperial College London, UK

W🌐 World Scientific

NEW JERSEY · LONDON · SINGAPORE · BEIJING · SHANGHAI · HONG KONG · TAIPEI · CHENNAI · TOKYO

Published by

World Scientific Publishing Europe Ltd.

57 Shelton Street, Covent Garden, London WC2H 9HE

Head office: 5 Toh Tuck Link, Singapore 596224

USA office: 27 Warren Street, Suite 401-402, Hackensack, NJ 07601

Library of Congress Cataloging-in-Publication Data

Names: Aliabadi, M. H., editor. | Sharif Khodaei, Z. (Zahra), editor.
Title: Structural health monitoring for advanced composite structures /
     edited by M. H. Ferri Aliabadi (Imperial College London, UK),
     Z. Sharif Khodaei (Imperial College London, UK).
Description: [Hoboken] New Jersey : World Scientific, [2017] |
     Series: Computational and experimental methods in structures ; volume 8 |
     Includes bibliographical references and index.
Identifiers: LCCN 2017023319 | ISBN 9781786343925 (hardcover : alk. paper)
Subjects: LCSH: Structural health monitoring. | Composite materials--Testing. |
     Structural analysis (Engineering)
Classification: LCC TA656.6 .S76 2017 | DDC 624.1/71--dc23
LC record available at https://lccn.loc.gov/2017023319

British Library Cataloguing-in-Publication Data
A catalogue record for this book is available from the British Library.

Desk Editors: Anthony Alexander/Jennifer Brough/Koe Shi Ying

Typeset by Stallion Press
Email: enquiries@stallionpress.com

Preface

Structural health monitoring (SHM) is a modern technology with the potential to significantly improve damage detection in composites and therefore act as a viable alternative to the commonly utilized non-destructive inspection (NDI). The industry's quest for high efficiency and performance with reduced weight has resulted in the extensive use of composite materials. However, composite materials are sensitive to in-service impact damage and with today's NDI technologies, inspections are frequent and costly due to the fact that it can only be carried out if access to the area to be inspected exists or is provided during maintenance checks.

The basis of SHM is the application of permanent fixed sensors on the structure combined with the necessity of a minimum of manual intervention to monitor the structural integrity. This enables a continuous monitoring of the structure, and thus a detection of the defect at a very early stage to move away from planned maintenance and towards condition-based maintenance.

This book, for the first time, provides an overview of prominent SHM techniques for damage detection and localization utilizing ultrasonic-guided waves in composites. It covers not only the fundamental concepts in the SHM piezoelectric and fiber optic sensor philosophy but also the state-of-the-art on passive and active sensing methodologies.

Chapter 1 initially provides a comprehensive review of guided wave damage detection techniques, before presenting some of the key aspects required for establishing an effective SHM system for large complex composites structures such as optimization of sensor positioning and influence of changes in environmental conditions on damage detection algorithms. Chapters 2 and 3 present efficient and accurate numerical modeling techniques for wave propagation and their interaction with damage in composites. Electro-mechanical impedance and guided wave propagation methodologies based on special signal processing techniques that allow identification of signal anomalies caused by structure degradation are the subject of Chapter 4. In Chapter 5, constructive interference via beamforming is developed to increase the reliability of the SHM system with a phased array configuration. Chapter 6 presents the fiber optic (FO) sensors as effective sensors for strain monitoring due to their high sensitivity to strain measurement, light weight and immunity from electro-mechanical interference. Finally, in Chapter 7, application of SHM to in-service monitoring of impact events is presented through passive sensing algorithms. Particular attention is paid to data-driven methods using machine learning algorithm for determination of impact location and energy. A Bayesian-based optimization is presented for determining the optimal sensor configuration in complex structures.

About the Editors

M. H. Ferri Aliabadi is a Professor or Aerostructures and Zaharoff Professor of Aviation. He has been the Head of Aeronautics Department at Imperial College, London, since 2008. Prior to joining Imperial College in 2005, he was Professor of Computational Mechanics and the Director of Aerospace Engineering at Queen Mary, University of London (1997–2004) and Reader and Head of Damage Tolerance Division at WIT, Southampton (1987–1997). Since 2004 he is the head of the Department of Aeronautics at Imperial College, London.

Ferri's main research interests are computational methods and mathematical modeling, aerospace materials and structures, composites and material innovations and structural health monitoring. He has published more than 250 journal papers and authored and edited over 50 books.

Zahra Sharif Khodaei is a Senior Lecturer in Aerostructures. She obtained her PhD from Czech Technical University in Prague in numerical modeling of functionally graded materials in 2008. Prior to her lectureship post in 2014, she was a Research Associate at the Department of Aeronautics, Imperial College, London, since 2009 where she conducted research in fatigue modeling and analysis of metallic and fiber metallic laminates (FMLs) and more significantly in developments of technologies and methodologies for structural health monitoring (SHM) of composite structures. Her main areas of research are structural integrity and health monitoring of structures in particular composites. She has authored more than 30 journal publications including book chapters.

Contents

Chapter 1

Damage Detection and Characterization
with Piezoelectric Transducers: Active Sensing

Z. Sharif Khodaei* and M. H. Ferri Aliabadi[†]

Department of Aeronautics, Imperial College London, London SW7 2AZ, UK
*z.sharifkhodaei@imperial.ac.uk
[†] *m.h.aliabadi@imperial.ac.uk*

This chapter presents an overview of prominent structural health monitoring (SHM) techniques for damage detection and localization utilizing ultrasonic guided waves (UGWs). The basic principles of SHM described include the application of permanently fixed sensors on the structure combined with the necessity of minimum manual intervention to monitor the structural integrity. The techniques used in SHM, especially tomography and delay-and-sum (DAS) approach, are described in detail. Particular attention is paid to the development of advanced technical capabilities for making the integration of sensors in modern composite structures practical and efficient so as to facilitate industrialization and certification. Therefore, key aspects that have been included are optimization of sensor positioning and influence of changes in environmental conditions on damage detection algorithms. The theoretical descriptions are combined with several benchmark examples involving finite element analysis and experimental measurements. Finally, application to a curved fuselage composite panel with frames and stringers is presented to demonstrate how a multilevel approach could be used to efficiently detect damage in complex structures.

1. Introduction

For designing modern materials for aerospace structures, a balance between low weight and high safety is required. Modern aircraft design typically follows damage-tolerant design concept. Nevertheless, the presence of damage in the structure can jeopardize the operation and

---

*Corresponding author.

safety if not detected and repaired at the right time. The recognition that safety, integrity and low weight are key priorities in the design of airframe structures has to led to developing techniques for monitoring and maintenance of composite structures. Since composites are relatively new materials in aerospace structures, the conventional non-destructive examination (NDE) techniques, which have been optimized for the needs of metallic structures, cannot satisfy the requirements for the maintenance of composite materials with a high probability of detection (PoD). This has led to a rather conservative design of composites applying strict damage tolerance rules, i.e., no growth rule.[1]

As the detection of barely visible impact damages (BVIDs) currently could not be guaranteed, damage tolerance design requires the composite structures to be rather thick, so that impact events do not reduce the strength below the limit load (LL) (see regulation ACJ 25.603). Depending on the adopted philosophy for the design and maintenance of future generation aircraft, one can decide either to establish frequent inspections, which mean a probable early removal of defect from service, or select a conservative design. Indeed, even when a conservative design is chosen, the detection of any damage, for instance during a walk-around, requires certain follow-up measures to characterize the damage further, and to ensure the airworthiness of the aircraft. These measures are especially costly when they are unscheduled, either due to the fact that they are accompanied by an unplanned down-time for the aircraft or the necessity to call for certified non-destructive inspection (NDI) personnel. Some estimates have claimed that 27% of average aircraft's life cycle cost is spent on inspection and repair.

Structural health monitoring (SHM) is an emerging technology covering the development and implementation of technologies and methodologies for monitoring, inspection and damage assessment based on integrated sensors. The acquired data in combination with advanced signal processing techniques can provide maintenance actions on demand. The continuous monitoring of the aircraft can result in condition-based maintenance (CBM), reducing the ground time and the maintenance costs significantly.[2]

There are various damage detection methodologies based on the type of analysis and type of sensors integrated in the structure.[3,4] The most popular sensor technologies are fiber optic (FO) and piezoelectric (PZT) sensors.

FOs can be embedded or surface mounted on the structure and can be used as strain measurement sensors either for damage detection[5,6] and/or load monitoring.[7–9] FO sensors are passive sensors. However, they can be used in combination with PZT actuators to form an active sensing network: a hybrid system. In the hybrid system, PZT actuators are used to excite the structure and FOs act as sensors to record the strain waves and can be used to monitor the state of composite structures.[10–12] FOs are an attractive sensor solution due to their light weight, high sensitivity to strain and ability to compensate for temperature effects. PZT transducers are used as actuators for exciting ultrasonic guided waves (UGWs) as well as sensing the propagating wave owing to their electro-mechanical coupling. They are an attractive solution due to their light weight, small size and low energy consumption. PZT transducers can be used for damage detection by actuating and sensing UGWs in the structure,[13–16] in vibration analysis[17–19] or by measuring the electro-mechanical impedance (EMI) response of the structure.[20–22]

An overview of the fundamentals of UGWs will be given in this section followed by the methodologies developed based on guided waves for damage detection and characterization.

Once an SHM system is designed based on the sensor technology and damage detection methodology, the decision to have a permanently installed sensor network for structural prognosis will be driven by its reliability, cost and the added weight of the system. The optimal placement of sensors/actuators in order to detect, with high probability and reliability, any damage before it becomes critical is a key factor in uptake of any SHM system. The interference of the sensor system with the design of the aeronautical part is required to be minimum. On the other hand, the SHM system must be able to detect various probable damage scenarios with high reliability and PoD. Therefore, optimization analysis needs to be carried out to find the best sensor layup (number and location) while minimizing the additional cost and weight of the system.

In this chapter, an overview of damage detection methodologies based on UGWs will be provided. The fundamental concepts of wave propagation together with key parameters in the design of any UGW-based SHM system is described in Section 2.1. Since the developed SHM system must work under the operational conditions of an aircraft, the effect of parameters such as temperature, humidity and vibration on the guided waves is also described. Different methodologies for damage detection based on sensor data are described in Section 3 and some of the most popular methods are further detailed and assessed. Once the damage detection methodology is developed, the next important question is how many sensors are required and where they should be located, which leads us to the optimization techniques outlined in Section 4. Finally, the applicability of the methodologies is assessed and validated by experiments on a large composite stiffened panel.

2. Overview of Ultrasonic Guided Wave Damage Detection

In this section, an overview of the basic principles of Lamb wave propagation in solids in the context of their application to NDE is outlined.

2.1. *Fundamentals of Guided Waves*

In general, elastic waves in solid materials are guided by the surface of the media in which they propagate. Rayleigh wave, defined as a surface wave, exists along the free surface of a semi-infinite solid decaying exponentially in its magnitude of displacement depending upon distance from the surface. However, in thin plate-like medium (such as aerospace panels), they are guided by the free upper and lower surfaces and are called guided waves. If the range of excitation frequencies is above 20 kHz, they are called ultrasound. The UGWs cause three types of motion in an elastic solid: waves polarized in the plane perpendicular to the plate are compressional (often called extensional) and shear (often called flexural) and waves polarized in the plane of the plate are called shear horizontal (SH) waves.

The governing differential equation of motion in an elastic filed can be written as

$$\mu u_{i,jj} + (\lambda + \mu)u_{i,ji} + \rho f_i = \rho \ddot{u}_i \quad (i, j = 1, 2, 3), \tag{1}$$

where u_i and f_i are displacement and body forces in the ith direction respectively, $\ddot{u}$ is the acceleration, ρ and μ are the density and shear modulus of the plate respectively and λ is the Lame constant. Equation (1) can be decomposed into two uncoupled parts under the plane strain condition, i.e., no SH waves (using Helmholtz decomposition[23]):

$$\frac{\partial^2 \phi}{\partial x^2} + \frac{\partial^2 \phi}{\partial z^2} = \frac{1}{c_L^2}\frac{\partial^2 \phi}{\partial t^2} \quad \text{governing longitudinal wave mode,}$$

$$\frac{\partial^2 \psi}{\partial x^2} + \frac{\partial^2 \psi}{\partial z^2} = \frac{1}{c_T^2}\frac{\partial^2 \psi}{\partial t^2} \quad \text{governing transverse wave modes,} \tag{2}$$

where ϕ and ψ are two potential functions defined as

$$\phi = [A_1 \sin (pz) + A_2 \cos (pz)] \, e^{i(kx - \omega t)},$$
$$\psi = [B_1 \sin (qz) + B_2 \cos (qz)] \, e^{i(kx - \omega t)}, \tag{3}$$
$$p^2 = \frac{\omega^2}{c_L^2} - k^2, \quad q^2 = \frac{\omega^2}{c_T^2} - k^2, \quad k = \frac{2\pi}{\lambda}.$$

Here, A_1, A_2, B_1 and B_2 are four constants determined by the boundary conditions while k, ω and λ are the wavenumber, circular frequency and wavelength of the wave respectively. Note that x is the wave propagation direction and z is the normal direction. Infinite wave modes are available in a finite body superimposing on each other between the upper and lower surfaces of the plate, and their propagation characteristics vary with entry angle, frequency and structural geometry. The longitudinal and transverse/shear modes propagate with different velocities denoted by c_L and c_T respectively and are defined by:

$$c_L = \sqrt{\frac{2\mu(1 - \nu)}{\rho(1 - 2\nu)}}, \quad c_T = \sqrt{\frac{\mu}{\rho}}, \tag{4}$$

where ν is the Poisson's ratio. Considering harmonic solutions for the potential functions in Eq. (3) and solving the wave equation and applying the appropriate boundary conditions at the upper and lower surfaces of the plate, we arrive at displacement fields representing general description of the Lamb wave in an isotropic and homogenous plate (with thickness $2h$)

$$\frac{\tan (qh)}{\tan (ph)} = \frac{4k^2 q p \mu}{(\lambda k^2 + \lambda p^2 + 2\mu p^2)(k^2 - q^2)}. \tag{5}$$

The tangent function in Eq. (5) (which can be defined with sine and cosine) has both symmetric and anti-symmetric properties. Considering the particle motion in the plate with respect to its mid-plane (xy plane), the above equation can be split into two parts with solely symmetric and anti-symmetric properties respectively, resulting in symmetric and anti-symmetric Lamb wave modes (see Fig. 1).

The two modes are denoted by the symbols S_i and A_i ($i = 0, 1, \ldots$) with the subscript being the order of the mode and S_0 and A_0 being the lowest (first) symmetric and anti-symmetric wave modes. S_i modes predominantly have radial in-plane displacement while A_i modes mostly have out-of-plane displacement.

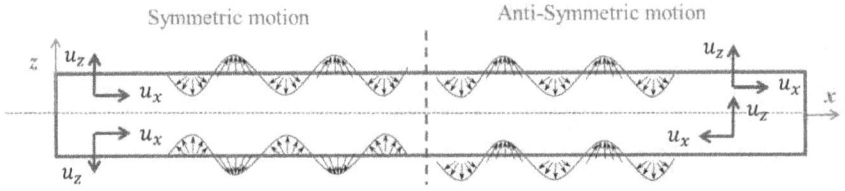

Figure 1.   Symmetric and anti-symmetric Lamb wave modes.

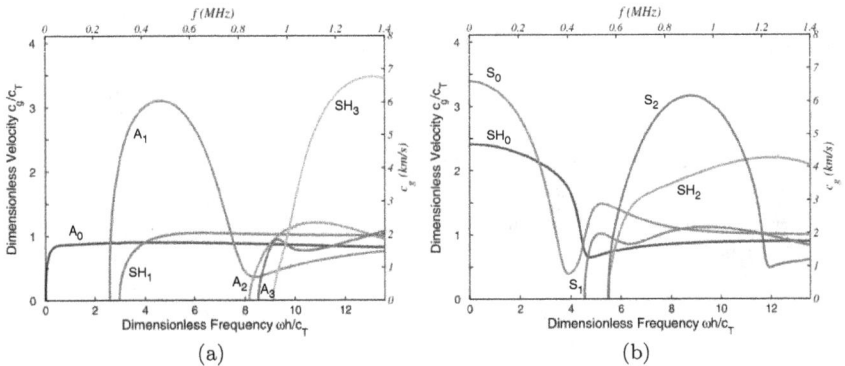

(a)

(b)

Figure 2.   Dispersion curves of Lamb waves along 30° direction in the laminate $[+45_6/-45_6]_s$. (a) c_g of anti-symmetric modes and (b) c_g of symmetric modes.[24]

Each wave mode has a different propagational velocity. The propagation of Lamb waves can be characterized by its phase and group velocities. The phase velocity is the propagation speed of the wave phase of a particular frequency contained in the overall signal and is calculated as $c_p = (\omega/2\pi)\lambda$. The velocity with which the overall shape of the amplitude of the waves (envelope) propagates through space is group velocity, c_g. The group velocity depends upon frequency and plate thickness. This phenomenon is referred to as *dispersion* and can be depicted graphically by dispersion curves as shown in Fig. 2. The dispersion curves are the solutions to the dispersion equation of Lamb waves (based on Eq. (5)), predicting the relationship among frequency, phase/group velocity and thickness.

For the application of the guided waves in SHM, the knowledge of the propagation velocity of the waves is very important. Therefore, it is often valuable to have dispersion curves. However, solving the dispersion equations of Lamb waves in laminated composite is not an easy task as

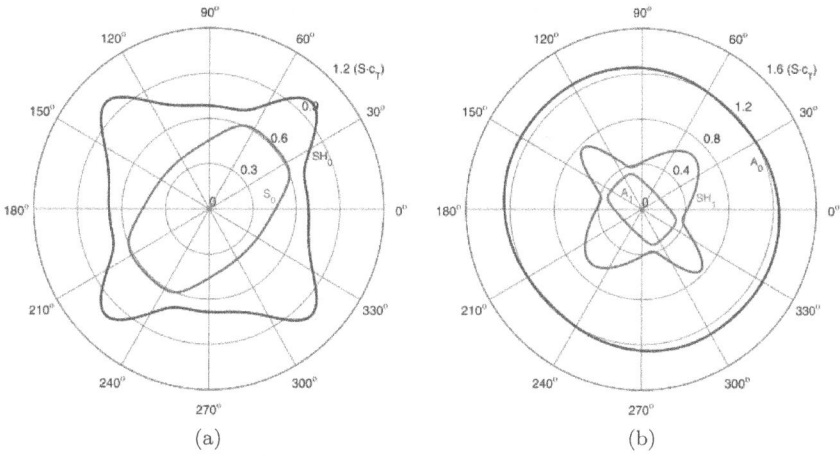

Figure 3.   Slowness curve of Lamb wave in the laminate $[+45_6/-45_6]_s$: (a) symmetric modes and (b) anti-symmetric modes.[24]

they do not generally have analytical solutions and in most cases can only be solved numerically or semi-analytically.[23,25–27]

In isotropic plates, the Lamb wavefront is circular and travels omni-directionally with the same velocity. However, in anisotropic materials, the wave velocity for each mode is subject to the direction of propagation. The directional velocity of Lamb waves for each propagating mode can be represented graphically with a slowness profile, which plots the propagation velocity as a function of propagation angle (see, for example, Fig. 3).

It is well known that energy dissipates with distance. This phenomenon also applies to Lamb waves as the gradual reduction in magnitude of wave signals and is known as attenuation. The presence of damage or even inhomogeneity in the structure such as stiffeners and fasteners increase the dissipation. This effect is particularly important in optimal placement of the sensors to cover large areas with high probability of detecting damage, which is related to the amplitude of the damage-reflected wave. Therefore, for every structure, the attenuation profile is required for the design of a reliable SHM system. It is worth mentioning that the S_0 mode tends to travel further than A_0 mode. This is attributed to the dominant out-of-plane displacements that leak partial energy to the surrounding environment whereas the symmetric modes mostly have in-plane displacements, which means that the energy will be confined within the plate. In addition, crossing stiffeners can introduce considerable wave scattering and energy dissipation in the structure, which needs to be taken into account in the design of the SHM system.

2.2. Influence of Temperature

As the temperature changes, it can result in changes in the material properties of the composite, and it can also affect the propagation of Lamb waves in three ways: length dilation changing the propagation distances, thickness dilation changing the frequency thickness product and changes in the material properties resulting in varying propagation velocities.[28]

The length and thickness dilation are result of density ρ changing with temperature T. For an isotropic material, this change can be measured by the equation $\rho = \rho_0(1 - 3\alpha\Delta T)$ There are two temperature-dependent material properties which lead to changes in the propagation velocity of Lamb wave with temperature: Young's modulus and Poisson's ratio (see Eq. (4)). The magnitude of Lamb wave signals increases linearly with temperature while the velocity decreases.[29,30]

Other studies on long-term effects of varying ambient temperature on Lamb waves have shown that there is a slight shift in the central frequency of the captured signals with regards to the excitation frequency.[28] The reason for this is changes in the PZT properties of the PZT transducers due to variation in the working temperature.[31,32] However, the change in the propagating wave in an isotropic structure such as aluminum is very different from an anisotropic composite. An experiment was carried out to study the effect of temperature on the propagating wave in an aluminum and a Carbon Fiber Reinforced Composite (CFRP) unidirectional composite plate with PZT transducers surface mounted on them.[33]

Figure 4(a) shows the temperature profile, in which both specimens have gone through six cycles while the wave propagation has been recorded. Figure 4(b) shows the peak of the signals recorded at various temperatures to highlight the amplitude reduction and phase shift due to temperature variations. These changes result in different measured Time of Arrival (ToA). The change in the arrival time is best highlighted by the short-time cross-correlation.[34] This change in aluminum plates is linear (see Fig. 5). However, when applying the short-time cross-correlation to signals recorded at varying temperatures in the composite plate, the behavior is no longer linearly increasing and it varies with the fiber directions (see Fig. 6). The impact on ToA is most significant for the wave propagating transverse to the fiber direction. This is due the stiffness of the plate in the transverse

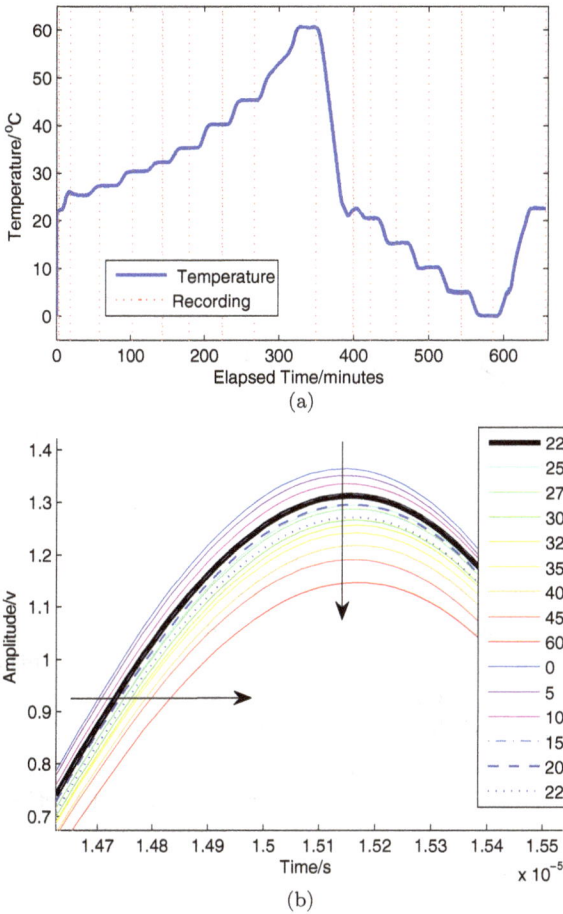

Figure 4. Temperature profile (a) and evolution of amplitude and arrival times (b) with change in temperature in an aluminum plate.[33]

direction being dominated by the properties of the resin, which reduces by 12% between the room temperature and 120°C.

   In addition, the change in properties of adhesive layer with temperature, different thermal expansion coefficients between the PZT and the host and slight change in the dimensions of the structure are additional factors for the slight shift in the central frequency.

   Since most of the damage detection algorithm based on UGWs, on comparing the actual state of the structure to the baseline measurements, when no damage has been present, the changes in the acquired data that are

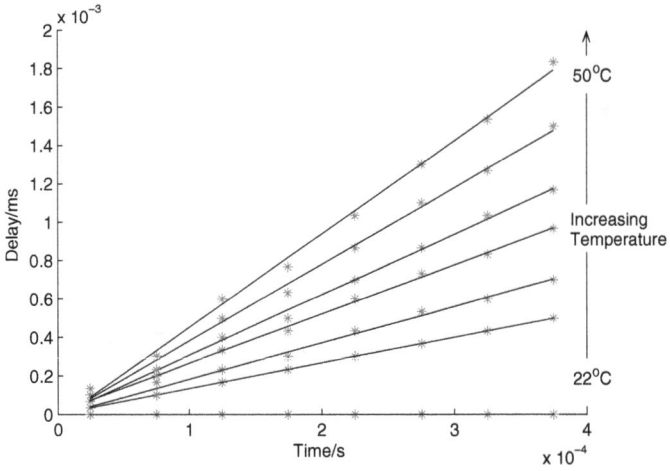

Figure 5.   Effect of temperature on ToA: Cross-correlation in aluminum plate.[33]

due to temperature variation can be mistaken for damage if not corrected. If the change in ambient temperature is small, there is no need to apply special corrections to compensate for the change. However, if the damage to be detected is small, the influence of the temperature may not be ignored as the modulation of the signal by damage may be overwhelmed by the changes associated with the change in temperature. Some of the developed temperature compensation techniques are presented in Section 3.2.

3. Methodologies for Damage Detection

UGW-based techniques are among the most effective methods for active sensing in plate-like structures owing to their sensitivity to small defects, low attenuation and large scanning areas. Once UGWs are actuated, their propagation properties depend on the media through which they travel. The presence of damage or defect can alter their propagation and allow damage detection and characterization when a sensor network is used. The sensor number and layout can then be optimized[35] to result in a system with high PoD.

Depending on the configurations, transducer can be used in pulse-echo or pitch-catch[36] mode (see Fig. 7). In the pitch-catch configuration, the sensors are placed in a network pattern and the structure is interrogated in

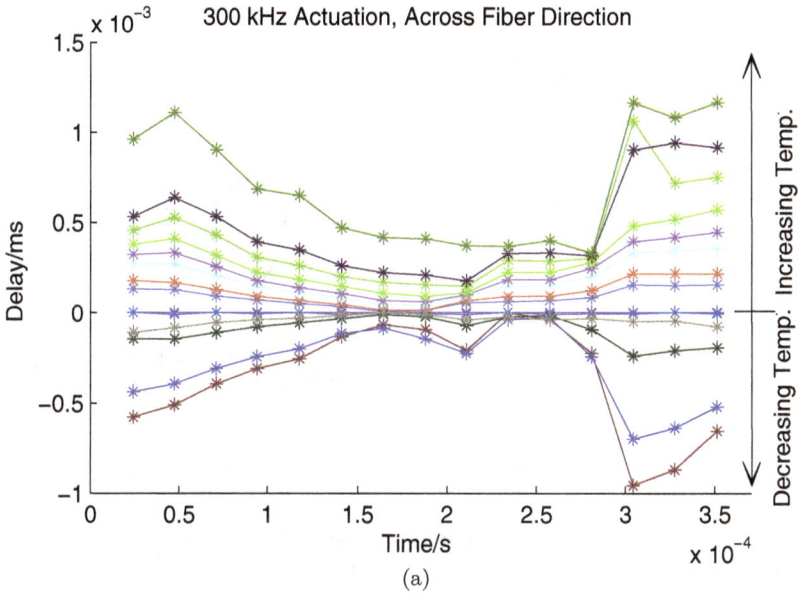

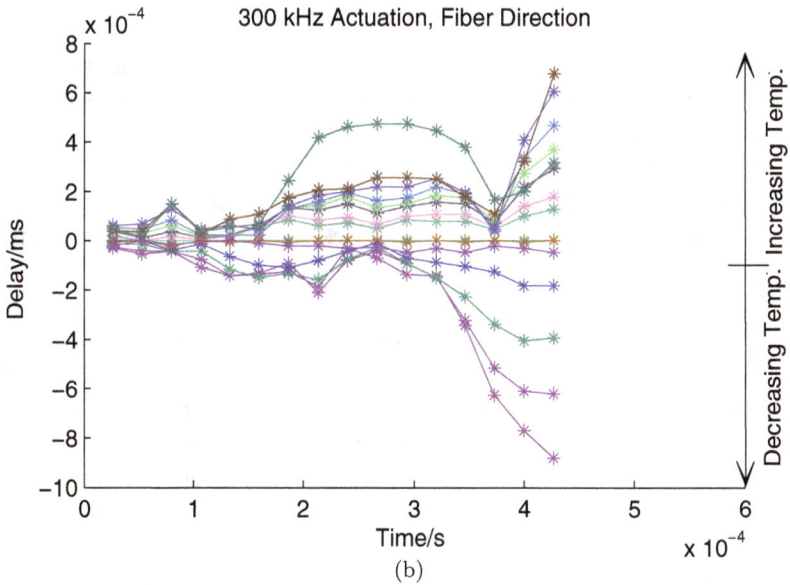

Figure 6. Effect of temperature on ToA: Cross-correlation in UD CFRP composite. (a) Shows transverse to the fiber direction and (b) shows along the fiber direction.[33]

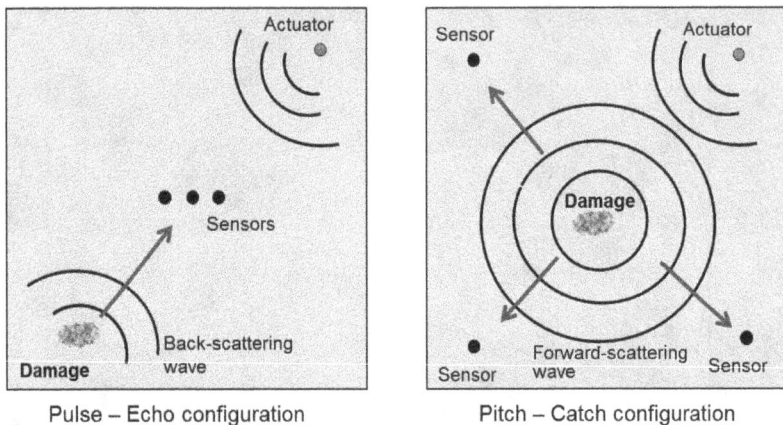

Pulse – Echo configuration   Pitch – Catch configuration

Figure 7.   Two possible operation modes for guided wave-based SHM.

a round robin approach. The damage is located inside the sensor network and is detected by capturing the forward scattered wave recorded by the sensors. In the pulse-echo configuration, the damage is located outside of the sensor network and is detected by capturing the back-scattered wave from the damage zone. The pulse-echo configuration represents phased array sensor configuration where each transducer can be used as a sensor and an actuator to excite the structure with a wavefront and record the reflected waves. One of the disadvantages of this configuration is that there will be a blind zone around the transducers, and for complex structures might not be easy to distinguish between the back-scattered signal from complex boundaries and those from the presence of damage.

There are different UGW-based techniques that require "baseline" signals related to the pristine state. The new measurements from the current state of the structure (possibly from a faulty state) are compared with the baseline signals and any significant change above a set threshold indicates damage. Often, the difference between the baseline response and the current response is called the residual. Section 3.1 presents different methodologies where a baseline response of the structure in pristine state is required. The focus of this chapter is on pitch-catch sensor configurations.

"Baseline-free" techniques are interesting since obtaining a reliable baseline in practice is challenging due to environmental effects, manufacturing processes and operational conditions. However, the techniques are less mature and mostly cannot deal effectively with complex structures. The most prominent baseline-free methodologies are summarized in Section 3.2.

3.1. *UGW Damage Detection Methodologies:*
Baseline Methods

The most straightforward method in obtaining the residual response is to subtract the pristine signals from the signals acquired at the inspection time and to apply a triangulation technique based on the difference in the time-of-flight (ToF) of signals to locate the damage.[37,38] The difference in ToFs is defined as the time difference between the incident wave that the sensor first captures and the subsequent damage (scattered wave). This suggests the relative position of the damage from each sensor–actuator pair. Considering a single sensor–actuator pair located at (x_S, y_S) and (x_A, y_A), the incident wave is defined as the first arrival of the signal through the direct path (d_{AS}) as shown in Fig. 8. Assuming that damage is centered at (x_D, y_D), the ToF of the damage scattered wave can be measured as

$$\Delta t = t_{ADS} - t_{AS} = \left(\frac{d_{AD}}{c_{g,1}} + \frac{d_{DS}}{c_{g,2}} \right) - \frac{d_{AS}}{c_{g,1}}, \tag{6}$$

where d_{AS}, d_{AD} and d_{DS} are the distances indicated in Fig. 8 and are measured by

$$d_{AD} = \sqrt{(x_D - x_A)^2 + (y_D - y_A)^2},$$
$$d_{AS} = \sqrt{(x_A - x_S)^2 + (y_A - y_S)^2}, \tag{7}$$
$$d_{DS} = \sqrt{(x_D - x_S)^2 + (y_D - y_S)^2}.$$

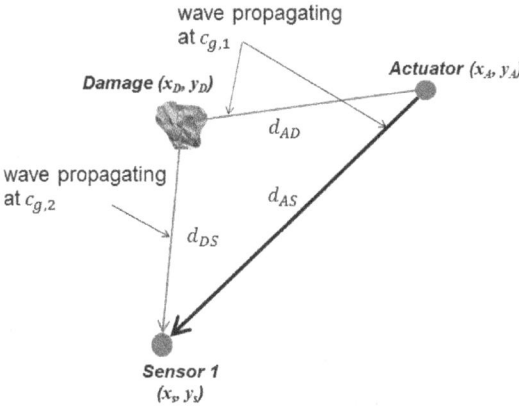

Figure 8.   ToF-based time of flight (ToF) triangulation: single path.

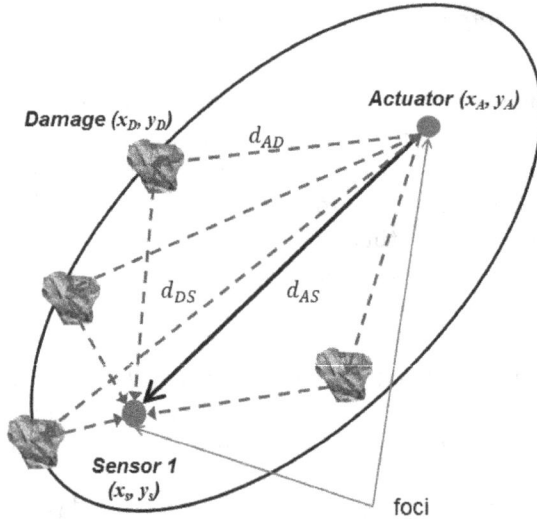

Figure 9.   Locus established by a sensing path based of ToF of damage scattered wave.

Furthermore, $c_{g,1}$ is the group velocity of the propagating wave from actuator to sensors and from actuator to damage while $c_{g,2}$ is the group velocity of the damage scattered wave received by the sensor.

Theoretically the solution to Eq. (6) configures a locus indicating all possible locations of the center of damage with the same ToF of the incident wave, as shown in Fig. 9. ToF extracted from other sensor–actuator pairs will also result in an equation similar to Eq. (6), which gives a group of nonlinear equations to be solved, which then result in intersection of different curves that provide the location of the damage center.

Mathematically, if $c_{g,1}$ and, $c_{g,2}$ are the same, the locus defined by Eq. (6) is an ellipse with the sensor and the actuator located at its foci. However, as the wave propagates through the damage area, mode conversion could occur upon interaction with damage and experimentally it will not be possible to know the magnitude of $c_{g,2}$ as it can vary for each damage scenario. Therefore, it is very important that the excitation frequency of the Lamb wave is within the non-dispersive zone, which means that if a mode conversion occurs after interaction with the damage, the group velocity will not change with a shift in the frequency ($c_{g,1} = c_{g,2}$).

This ToF-based triangulation is effective for simple and small structures. However, for a complex structure such as a stiffened panel, where stiffeners and frames cause additional reflections, the analysis of Lamb waves becomes increasingly difficult.

The residual must only contain damage-scattered signal. If the baseline is recorded at a different condition to that of the current interrogation response, the accuracy of damage detection and localization can be significantly reduced.[39] Environmental factors such as temperature rise reduces propagation velocity,[40] resulting in non-damage-related residual, which, if not corrected, results in a false alarm. Temperature correction methods compensate for changes in the velocity using baselines recorded at a range of temperatures.[41] Subsequently the optimal baseline signal can be selected and corrected, for example, by stretching the signal (baseline signal stretch BSS). The BSS method can be performed in time or frequency domain, both with a similar level of performance.[42,43] An improved BSS method was proposed.[44] where different correction factors was used for each wave packet and it was shown that the baseline correction was valid for a wider range of temperature changes and applicable to both metallic and composite structures. However, to obtain this range of data is very time-consuming and if any changes are made to the structure (operational or mechanical), the entire baseline library has to be repeated.

The sensor signals in their discrete form carry too much information to be used directly for damage characterization. Consequently, different identification algorithms are developed based on a range of feature extraction techniques. A comprehensive selection of methodologies are reported in Ref. [45]. Different signal features can be used to define a damage index for localizing and characterizing a damage. For an array of transducers, phased array transducers distributed in a ring pattern can be used to detect and localize damage.[46] This method has been shown to be effective for damage localization, by transforming the signals from their time domain to their spatial domain; nonetheless a large number of sensors in close vicinity of the damage was necessary (12 sensors in a circular area of 80 mm diameter), which makes it an attractive method for localized searches in structural hotspots. A second algorithm based on constructing a damage influence map of the structure was tested, which is based on comparison of the baseline and current signals at defined time windows (in a similar manner to delay-and-sum (DAS) algorithm).[47] Spectral elements were used to model the wave propagation in a multilayer composite and cracks in the structure were successfully detected using the numerical model. However, the applicability of most of the proposed methods to complex composite structures under operational load with real impact damage is questionable as they have been developed for simple structures, requiring large number of sensors and in most scenarios, damage has been simulated by adding mass or making through thickness holes.

One of the challenges of the UGW methods applied to built-up structures is the complexity of the received signal even when there is no damage, consisting of multiple overlapping reflections from structural features. Baseline subtraction can get rid of the most of the coherent structural reflections, which may obscure the damage effects. One of the first methods based on signal subtraction from an array of transducers combined with an imaging algorithm to locate the damage is the DAS method proposed by Wang et al.[48] DAS is an imaging algorithm where the probability of having a damage located at every point in the structure is measured from the ToF of the scattered signals. Subsequently DAS method has been employed by many others[49,50] proposing modifications and improvements to reduce the effect of the boundary reflections.[51–53] There are several other imaging algorithms such as the minimum variance (MV) method[54] where damage is not only detected but also characterized by generating MV images for various scattering assumptions and determining which image contains the strongest response at the potential damage location. However, MV requires the knowledge of expected scattering patterns for all incident scattered angles, which is not feasible.

In a similar approach to DAS, Bayesian imaging is proposed where the most probable damage location is generated by measuring the likelihood that a scattered arrival occurs at a particular time, from all transducer pairs.[14] More recently the maximum likelihood estimation (MLE) approach was applied together with a proposed sensor fusion, based on Neyman–Pearson (NP) criterion, where the impact of sensor pairs not observing the damage is mitigated for the study in Ref. [55]. This method was successfully applied to complex steel structures.

Another group of UGW imaging techniques, which is well established along with DAS, is the tomography approach. The idea of guided wave tomography was initially inspired by the computational tomographic technique used in X-ray imaging. The most developed tomographic approach is the reconstruction algorithm for probabilistic inspection of defects (RAPID) algorithm proposed in Ref. [56] where the signal difference coefficients (SDCs) extracted from the guided wave signals is used as inputs to the image reconstruction algorithm and has been successfully applied to complex structures, such as aircraft wing parts.[57,58]

Both DAS and tomography approaches are widely used for applications to complex structures and therefore can be categorized as the most mature methodologies with the possibility of being applicable to real structure under operational conditions. They are further described in Sections 3.1.1 and 3.1.2 with some applications to complex structures presented in Section 5.

3.1.1. *Tomography Approach*

Tomography is an imaging approach based on the residual signals (difference between the pristine and the damage states). The presence of a defect is revealed in the tomograms constructed from the residual signals. A tomography-based algorithm called RAPID has been developed, which accounts for wave scattering and reflections from damage using a probabilistic damage detection concept by superposition of rays of ellipses.[58] An SDC to quantify the changes in the signal was proposed based on a correlation analysis

$$\rho = \frac{C_{BD}}{\sigma_X \sigma_Y}, \tag{8}$$

where C_{BD} is the covariance of the pristine data set B_{ij} and each new set D_{ij} recorded during the service time, σ_B and σ_D are the standard deviations of B_{ij} and D_{ij}.

$$C_{BD} = \sum_{k=1}^{n_s} \left(B_{ij}[k] - \mu_{ij}^B\right) \left(D_{ij}[k] - \mu_{ij}^D\right), \quad i, j = 1, \ldots, n, \tag{9}$$

where the index $[k]$ indicates the discretely sampled signal, n_s is the number of samples in signals B_{ij} and D_{ij}, and μ_{ij} denotes the mean value of the corresponding signal between actuator i and sensor j.

The location of the defect is determined by the severity of signal changes of different sensor pairs as a result of the defect. The location is then expressed as a probability of defect distribution by a linear summation of the correlation coefficients from all actuator–sensor pairs. If the sparse array of transducers is constructed of n transducers, then the total number of acquired signals is $n(n-1)$ The total number of signals is halved if the reciprocity of the system is assumed.

The tomography approach can be divided into two steps: (i) the SDC calculation and (ii) the data fusion and imaging. Once the SDC coefficients are calculated for all transducer pairs, an imaging algorithm will be applied. The spatial distribution of defect probability, s_{ij} on a direct path is assumed to be linearly decreasing elliptical distribution as shown in Fig. 10. The size

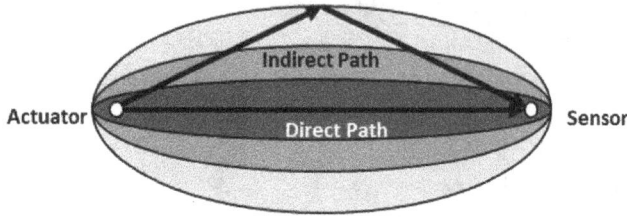

Figure 10.   Elliptical distribution function of probability of defect location.

of the elliptical distribution function is controlled by a scaling parameter β, which is usually selected to be around 1.05.[58] In general, when a defect occurs, the sensor signals will be affected and consequently the defect distribution probability image will have higher probability where the defect is located compared to other points estimated from the severity of the signal changes of different transducer pairs.

The physical reasoning behind this is that a defect would cause the most significant signal change in the direct path of the wave and the effect of signal change will decrease away from the direct path.

The defect probability distribution s_{ij}, for any point x, y is then defined as follows:

$$
s_{ij}(x,y) = \begin{cases} \dfrac{\beta - R_{ij}(x,y)}{\beta - 1} & \text{if } \beta > R_{ij}(x,y) \\ 0, & \text{if } \beta \leq R_{ij}(x,y), \end{cases} \tag{10}
$$

where $R_{ij}(x,y)$ is a geometrical function representing the ratio of the distances ($|$actuator to the point of interest$| + |$point of interest to sensor$|$) and $|$actuator to sensor$|$. β is a scaling parameter that controls the size of the effective elliptical distribution area shown in Fig. 10 (usually β is selected around 1.05).

The values of s_{ij} are evaluated for all transducer pairs and at all points (x, y) of the structure (divided into a grid). The defect distribution probability $P(x, y)$ is then expressed as fusion of the damage index from individual pairs as a linear summation

$$
P(x,y) = \sum_{i=1}^{n} \sum_{j=1}^{n} \text{SDC}_{ij} s_{ij}(x,y), \tag{11}
$$

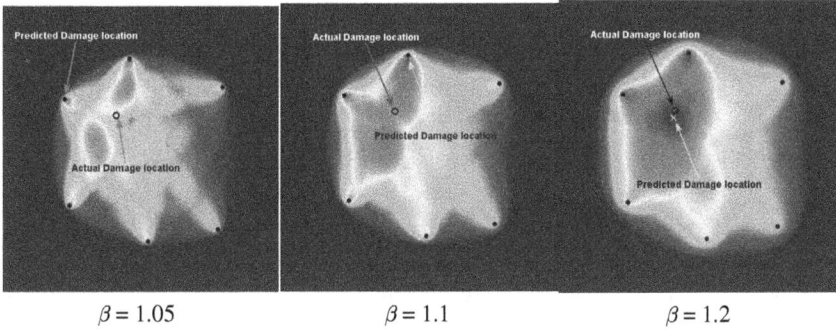

$\beta = 1.05$ $\beta = 1.1$ $\beta = 1.2$

Figure 11.   Damage detection using RAPID algorithm with different scaling parameter.

where the SDC_{ij} value quantifies the difference between each two signals (baseline and current states) and can be calculated using the correlation coefficient ρ defined in Eq. (8) for every transducer pair ij

$$SDC_{ij} = 1 - \rho_{ij}, \quad ij = 1, \dots, n. \tag{12}$$

The RAPID algorithm has been successfully applied for detecting single and multiple damages in composite structures.[59,60] However, most of the successful results including the multiple damage detection have been obtained from numerical models where the influence of the environmental factors and operational conditions were not present. In addition, for the experimental cases, a very dense sensor network is required to obtain reliable results.[59,61] This is not very attractive option for application to real structures due to the additional weight of the sensors and wires and the extra acquisition time.

The success of the RAPID algorithm depends on several parameters that must be considered prior to its application to complex structures. The first factor is the choice of the scaling factor β, which is chosen arbitrarily. Varying the scaling factor β from 1.05 to 1.2 results in very different diagnosis as shown Fig. 11. For example, the effect of the scaling factor was shown on an aluminum plate with through thickness hole.[51] In addition, when applied to detecting a damage under the stiffener, it was shown that when the number of transducers is not sufficiently high (four transducers for a composite plate of 300×225 mm), damage could not be detected with RAPID while DAS successfully detected the same damage with the same number of transducers.[51]

3.1.2. Delay-and-sum

DAS is an imaging algorithm for a sparse array of transducers where the damage is located by measuring the probability of damage from the residual response of the structure. The residual response of the structure isolates any changes between the two signals. For a guided wave imaging to result in reliable diagnosis, it is important that the residual corresponds to the scattering only from defects or damage. Under real operational conditions, however, there are a number of changes in the signals that are unrelated to damage including environmental conditions, surface conditions and operational loads. DAS is also called an elliptical method because the constant arrival time curve for a scattered echo is an ellipse with the transmitter and receiver located at the foci (see Fig. 9). Elliptical imaging is performed first by dividing the structure into pixels. The probability of damage being located at that pixel is then presented by the value of the damage index (DI) measured from the residual signal at a time of arrival of the scattered signal, t_{ADS}, as indicated in Fig. 12. Consider that damage is located at pixel location (x_D, y_D), then a residual signal corresponding to path ij will contain some scattered energy at time t_{ADS} defined as:

$$t_{ij}^{xy} = \frac{d_{ij}^{xy}}{c_g}, \tag{13}$$

where ij indicates a specific transducer pair, xy refers to the coordinates of the pixel (for example x_D, y_D shown in Fig. 9), d_{ij}^{xy} is the distance from the actuator to the pixel and then back to the sensor $(d_{AD} + d_{DS})$ and c_g is the group velocity.

The residual signal, as presented in Fig. 12, is the difference between the energy envelope between the signals from the pristine state and the current state. This ensures that the DI value at any time instance is always a positive measure and it is also a means of denoising the signals. If we assume that the difference between a pristine signal and a current state signal is denoted by $U_{ij}(t)$, the complex analytical signal $C_{ij}(t)$ is formed from the original signal and its Hilbert transform $V_{ij}(t)$

$$C_{ij}(t) = U_{ij}(t) + iV_{ij}(t). \tag{14}$$

Then, the envelope of the residual signal is the magnitude of the analytical signal

$$r_{ij}(t) = \sqrt{U_{ij}(t)^2 + V_{ij}(t)^2}, \tag{15}$$

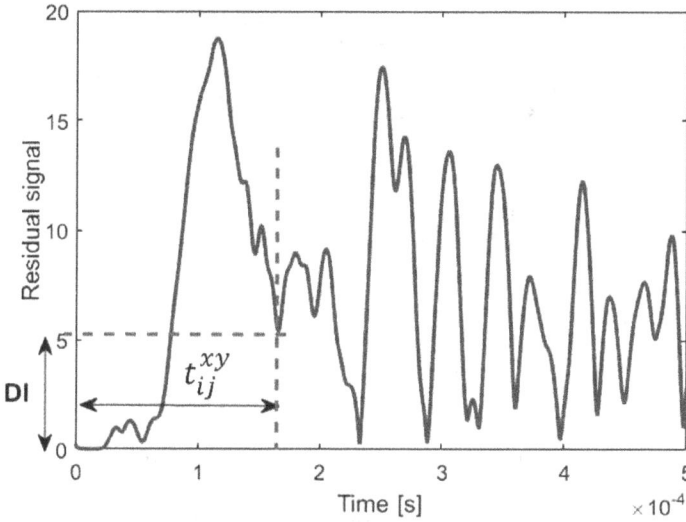

Figure 12.   Example of DI measured at one pixel based on ToF of damage scattered wave.

where n is the number of transducer pairs. Each differenced signal is then delayed, squared and averaged at each spatial location (x, y) following Eq. (16):

$$\text{DI}(x, y) = \frac{1}{n} \sum_{i=1}^{n-1} \sum_{j=i+1}^{n} r_{ij}(t_{ij}(x, y)). \tag{16}$$

It should be noted that the above equation is sometimes presented using an additional integration over a predetermined time window.[62] This will improve the imaging performance by reducing the integration window to an instantaneous point in time. Alternatively, the weighted energy arrival method (WEAM) proposes adding a weighting function $w_{ij}(\sigma, v)$ to the energy envelope in order to isolate the DI calculation to the first mode of the residual signal and avoid secondary reflections from boundaries and/or superposition of higher mode as shown in Fig. 13.[51] The envelope of the residual signal, r_{ij} is weighted by a log-normal distribution having the mean v centered at ToF of the first peak of the damage reflected wave

$$\text{DI}(x, y) = \frac{1}{n} \sum_{i=1}^{n} \sum_{j=1}^{n} r_{ij}(t_{ij}(x, y)).w_{ij}(\sigma, v), \tag{17}$$

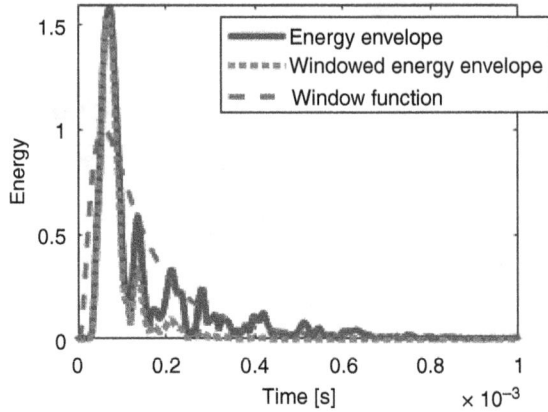

Figure 13.   Influence of the weighting function on the residual signal.

where $w_{ij}(\sigma, v)$ is a window function with log-normal distribution. Most of the proposed DAS methods assume reciprocity of the solution and therefore only use $n(n-1)/2$ paths in the above equation. This means that the recorded signal in sensor 2 from actuator 1 should be equal to signal in sensor 1 from actuator 2. This is true only when damage is not present in the structure since the damage will disturb the symmetry of the problem, or only when the damage is symmetric with respect to that path, which is not expected as a result of an impact on a composite plate.

To detect BVID experimentally, the result of the damage characterization (sizing) can be significantly improved when reciprocity is not used, in particular for a structure with minimum number of transducers where the paths go directly through the damage area[51] (see Fig. 14).

So far, different methodologies based on the difference between signals acquired at different states (pristine and current) have been presented. For large and complex structures, to ensure the reliability of the detection, the first step is to carry out a case study on the pristine panel to find the attenuation profile of the signals, in particular when the wave crosses stiffeners and frames. In addition, a reliable threshold value for the DI measures must be set. The attenuation profile is generated to ensure that the amplitude of the traveled wave is high enough to detect changes in the structure, which are due to the presence of damage and well above the

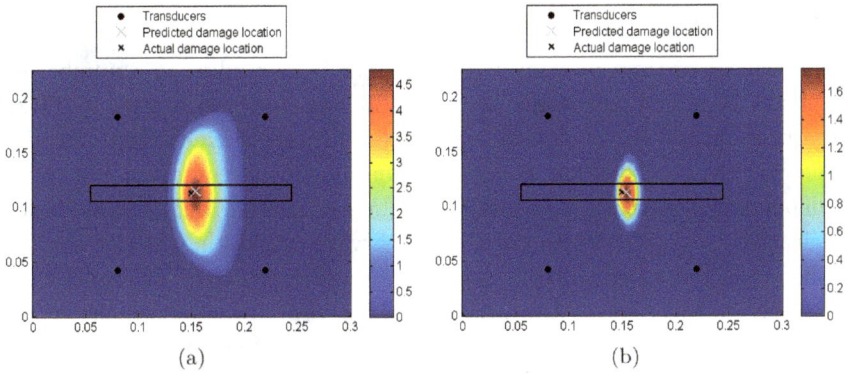

Figure 14. Improved DAS algorithm applied to the stiffened composite panel. DI measured from (a) reciprocity of paths and (b) all possible paths.

noise level. For the pristine structure, reciprocity of the paths must hold, i.e., the signals from path 1–2 must be identical to path 2–1. However, in real structures, there are tolerances with respect to installation of the transducers, which can result in small variation in identical paths. These small perturbations can be measured for the total sensor network and a detectability threshold set for the DI value. To define a damage threshold, usually a large library of baseline measures is required to carry out a statistical analysis to separate the environmental effects, geometrical factors and installation tolerances from the damage effects in the signals. Using this library, a reliable threshold must be defined, which indicates the existence of damage, along or very close to that transducer path, when the DI value is above this threshold. Some examples of the threshold measurement are presented in Section 5.

3.2. *UGW Damage Detection Methodologies: Baseline Free*

The ongoing efforts to minimize the false alarm and increase the reliability of the decision-making of an SHM system operating under varying operational and environmental conditions has led to methodologies that do not depend on previously obtained baseline data. These techniques can be categorized mainly into three approaches: (1) nonlinear detection approaches such as time-reversal method (TRM)[63]; (2) mode-conversion-based techniques[64] and (3) sensor-array-based techniques.[65]

3.2.1. *Time Reversal Method*

The concept of time-reversal invariances has been studied in a number of contexts including ultrasonic imaging and non-destructive testing. TRM has been proposed to mitigate the need for a baseline signal for the following reason: Since the governing equations for stress waves in a time-independent structure, considering no losses, contain only second-order derivatives with respect to time, for any wave originating from a source, either reflected or refracted by scatters, there exists a set of waves that precisely retraces all of the complex paths and converges synchrony at the origin of the source, as if the waves were propagating backwards in time.[66] In practice, the guided wave signals originated by an actuator, after going through all the multiple scattering, reflections and refractions, are recorded by an array of sensors, time-reversed and re-emitted back into the structure. The re-emitted signals propagate backwards through the media and refocus on the source. Based on the assumption of linear reciprocity within a healthy structure, the final signal and the original should compare exactly. However, when nonlinearities such as damage are introduced into the system, linear reciprocity is not valid anymore and the difference between the two signals indicate the presence of damage. For non-dispersive waves, when the scattered signals are time-reversed and re-emitted, both temporal compression and spatial focusing can be achieved. However, for dispersive waves such as Lamb wave, wave re-compression occurs at the source location, giving rise to high amplitude pulse. Theoretical and experimental investigation of the influence of dispersion of Lamb waves on temporal and spatial focusing of time-reversal waves in plate like structures was carried out.[48] Furthermore, a synthetic TRM was proposed to improve the signal-to-noise ratio. The method was reported to successfully detect a mass bonded to a plate, but it has not been tested for a damaged composite plate. TRM was applied to detect and locate delamination in composite plate using a rigorous statistical classifier based on extreme value statistics.[67] However, for a plate of size $60.96 \times 60.96\,\mathrm{cm}$, 16 transducers were used, which adds up to 240 different actuator–sensor pairs, which is very time-consuming and computationally expensive. A modified TRM was presented, which reduced the hardware requirements significantly and successfully detected the presence and severity of impact damage in a composite plate.[68] However, in complex structures, there could be other sources of non-linearity such as change in thickness in composite plates, stiffeners and bolts, which can be mistaken for damage.

Another recently developed nonlinear method is based on the tomography approach (i.e., RAPID algorithm)[60] introduced in Section 3.2.1 where an SDC parameter is introduced based on scaling subtraction method (SSM), which eliminates the requirements for an intact baseline.[69] The theory behind the nonlinear RAPID algorithm is that the defect behaves as a nonlinearity with increasing excitation amplitude. The excitation amplitude has to be kept sufficiently low in order not to provoke any nonlinearity in the transducer response. This limit can be obtained experimentally by direct monitoring of the output waveform distortion as a function of input amplitude. The steps of the nonlinear RAPID algorithm are as follows[60]:

(1) A low-amplitude excitation signal is applied to a sparse array of transducers and the corresponding response signals B_{ij} are obtained where they will be acting as reference signals.

(2) A high-amplitude excitation is applied under the same environmental conditions and the response signals D_{ij} are obtained.

(3) A nonlinear SDC coefficient is proposed by Eq. (18)

$$\text{SDC}_{ij} = \frac{1}{n} \sum_{m=1}^{n} \left(k_s B_{ij}(t_m) - D_{ij}(t_m) \right)^2, \tag{18}$$

where k_s is the amplitude difference between signals B_{ij} and D_{ij}.

The nonlinear RAPID was applied to both numerical and experimental signals and in both cases compared to the linear RAPID with baseline measures taken at a pristine state. Even though for the case of experimental results, the baseline free RAPID results have lower imaging quality and the detected shape of the damage distribution does not fully comply with the actual damage area, it is still an attractive method if it can be applied to more complex structures. However, the same disadvantages that were highlighted in previous section still apply here and the methodology requires further development and validation.

More recently, a baseline-free damage index method has been proposed based on node displacement measured from the modal analysis of the structure.[70] The idea of the proposed method is that in the mode shape, the displacement at the node is always zero. However, the change in the structural stiffness due to the presence of damage will cause a non-zero node displacement. This method was tested both numerically and experimentally on a steel beam. The damage was simulated by adding weights from 2 to 40 g. To convert the recorded data from time to frequency domain, fast Fourier transform was used. The magnitude of the node displacement will vary with the applied impact load. Therefore, a transfer function was proposed, which depends on the structural characteristics only. However, the applicability of the method needs to be demonstrated on composite structures and more complex geometries.

3.2.2. *Mode Conversion-based Techniques*

Once the guided wave propagates through a damage, mode conversion can occur depending on the characteristics of the damage. By identifying the mode conversion related to presence of damage, the damage in the structure can be detected. However, this technique requires the isolation of individual modes, which can be a very challenging task due to the existence of multiple reflection and overlap among different Lamb wave modes. To overcome this, a method was proposed based on the transferred impedances obtained between two pairs of collocated PZT patches.[71] This means that PZT transducers have to be installed from both sides of the structure and this often is not acceptable or possible in practice.

3.2.3. *Sensor Array Baseline-free Techniques*

The third group of baseline-free techniques requires an array of sensors. The concept behind the instantaneous baseline measurement is that transducers can be placed on a structure such that pitch-catch Lamb wave propagation can be used to obtain common features of undamaged sensor–actuator paths to act as a baseline. The transducers must be placed such that the sensor–actuator paths are of equal length and that structural features and material properties are spatially uniform between transducers,[65] which will not be applicable to complex, anisotropic structures. However, a more recent instantaneous baseline technique has been proposed for a large structure by dividing the structure to sub-sections based on the geometrical similarities and using sensor mapping[72] and applying delay and sum algorithm. Through sensor calibration, it is shown that the damage scatter signal can be extracted, allowing accurate localization using time of arrival and amplitude. This approach can be applied to complex structures with periodic features such as a composite stiffened panel by exploiting the symmetry (or partial symmetry) of the structure. The problem of environmental conditions and temperature change is mitigated as baselines are recorded at the same time as the current signal. The proposed instantaneous baseline method is detailed in the next section.

3.3. *Instantaneous Baseline Method*

The instantaneous baseline method presented in Ref. [72] generates calibrated baselines recorded at the same time as the current/damaged signal. For any interrogation path, a geometrically similar path is used as an instantaneous baseline recorded under the same condition and therefore mitigating the environmental effects such as temperature and humidity. Due to geometrical and manufacturing tolerances, no two pairs of transducers are expected to have the same exact recorded signals. Therefore, the signals are then calibrated to be used as baseline signals for input to any damage detection techniques such as DAS. The baseline path B_j is mapped onto the current interrogation path C_i. To minimize the mismatch of the paths, a calibration factor L_{ij} is proposed for each transducer path

$$L_{ij} = \text{env}\,(B_i - B_j)\,; \quad R_{ij} = \text{env}\,(C_i - B_j)\,;$$
$$R_{ij}^c = R_{ij} - L_{ij}, \tag{19}$$

where B_i is a geometrically similar path in the baseline area, for example path 4–5 shown in Fig. 15 for the interrogated path 3–6.

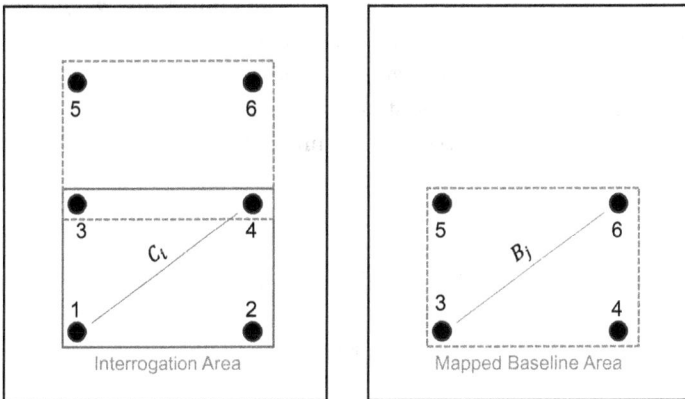

Figure 15.   Geometrically similar baseline path mapped onto inspection path.

For this method, the sparse transducer array must be arranged to form repeating cells. Even if the structure as a whole is not symmetric, the internal symmetry of the structure can be exploited such as a fuselage barrel, which can be divided into different bays. The application of the purposed method on a composite coupon has been reported in Ref. [72] and its application to a composite stiffened panel is validated and reported in Section 5.1.2.

4. Optimization

For aeronautical applications, the decision to have a permanently installed SHM system for structural prognosis will be driven by its reliability, cost and the added weight of the system. A challenge for guided wave-based methods, in particular for composite structures, is in separating mode-superposition and reflection of waves from boundaries and those emanating from damage (see e.g., Ref. [14]). This highlights the need for determining the ideal actuation signal and optimal sensor placement, particularly in complex structures such as aircraft composite panels with stiffeners, frames, rivet holes and large manholes.

Optimal placement of sensors/actuators in order to detect, with high probability and reliability, any damage before it becomes critical is a key factor in uptake of any SHM system. The position of sensors can have a significant influence on the values of the DI.[19] The effect of the pattern of the sensor layout on the detection capabilities of guided waves in metallic structures can be found in Refs. [73–75]. An optimization approach for damage detection was proposed with the objective function derived using artificial neural networks (ANNs).[76] This approach would require a significant number of damage scenarios to be used in the training of the network. Similarly, an improved genetic algorithm (GA) for optimal sensor placement for a metallic truss structure has been developed.[77] Another proposed optimization strategy is to minimize the Bays risk.[78,79] This methodology is based on a statistical model of the active sensing process with guided waves by deriving the signal properties such as scattering envelope due to a defect (in this case crack in a metallic structure). A similar approach incorporating minimum number of transducers with line of sight (LOS) was later developed,[80] which minimizes negative effects of boundary reflections and accounts for transducer failures under operational conditions.

In a more recent study, an optimal sensor placement algorithm based on GA is proposed[81] where the objective function is independent of the damage detection algorithm and is based on the geometry of the sensors and the PoD function. The PoD function can be constructed from the energy scattered profile of Lamb waves based on ray tracing approach where only two parameters are required: Lamb wave group velocity and their spatial attenuation.

All of the above-mentioned methodologies require an objective function, which merely relies on the performance of the damage detection methodology, i.e., PoD. This means that a vast number of damage scenarios (various locations and severities) have to be analyzed for each structure under investigation in order to determine the optimal sensor configuration. Alternatively, a recent study has proposed an optimization strategy based on maximum area coverage (MAC) within a sensor network.[82] The advantage of this approach is that it is independent of the details of the damage detection algorithm and does not require determination of a PoD function for a vast number of damage scenarios. Moreover, it can be applied to geometrically complex structures with pitch-catch sensor configuration and any active sensing procedure based on ToF of damage reflected waves, such as tomography approach or DAS. This method is described in Section 4.1.

4.1. *Optimal Transducer Arrangement: Maximum Area Coverage (MAC) Approach*

As described in earlier sections, operational loads and environmental conditions such as variations in humidity and temperature result in changes in amplitude and phase of the Lamb waves. Therefore, optimization methods that utilize PoD as their fitness function are unlikely to allow for all the above conditions together with geometrical complexities, which are present in real structures. Therefore, a robust optimization algorithm should allow for optimum sensor placement irrespective of the operational loads and environmental conditions as well as allowing for incorporation of specific structural complexities. To this end, an algorithm based on maximum coverage (MAC) was developed in Ref. [82] for optimal sensor placements for damage localization. The strategy was developed for the pitch-catch mode guided wave damage localization procedures. The fitness function is introduced as an indicative measure of probability of detecting damage at any point in the structure. This value is based on maximizing the coverage area within the sensor network while reducing the negative effects of the boundary reflections. The proposed fitness function is geometrical and physical based. To introduce a coverage map of the structure, the geometry is divided into pixels where the fitness function is evaluated at the center of each pixel. Increasing the number of pixels will increase the resolution of the fitness function. For every pixel, each transducer path resulting in constructing an ellipse passing through its center will contribute to the fitness function measure at that point, i.e., the contribution equals to one. This will then be carried out for each pixel and summed for each transducer path. The higher the number of ellipses passing through a single point, the higher the probability of damage being detected at that given location. Given a total of n transducers, the number of ellipses crossing a single pixel is given by $L = n(n - 1)$. The geometrical constraints that influence the fitness function are: weighting of the direct path (higher probability of damage being detected) and the negative effect of the reflections from the physical boundaries of the structures such as edges of an opening and stiffeners. The physical constraints that are built into the definition of the fitness function of MAC are the group velocity and the attenuation profile of the UGWs. The combination of the geometrical and physical factors then defines the global fitness function, which needs to be maximized in order to cover a larger area of the structure but at the same time reducing the negative influences of the boundaries. The optimization strategy is then based on GA to find the best sensor combination, which maximizes the probability of damage being detected inside the transducer network.

The intensity of the coverage can be visualized by a coverage map where the higher values refer to higher probabilities of damage being detected at that point. The boundary reflection constraints, γ, reduces the intensity of the coverage due to unwanted reflections. For a rectangular plate with four transducers, depending on the coefficient γ, the intensity of the coverage can change significantly; see Fig. 16 where the maximum intensity (and consequently the coverage) reduces by increasing the values of coefficient γ. The maximum intensity of the coverage is the total number of ellipse intersections obtained from all transducer pairs at each point plus the additional weight for the direct paths. Due to the boundary reflection constraints, for some points, there will be no contribution from certain transducer pairs.

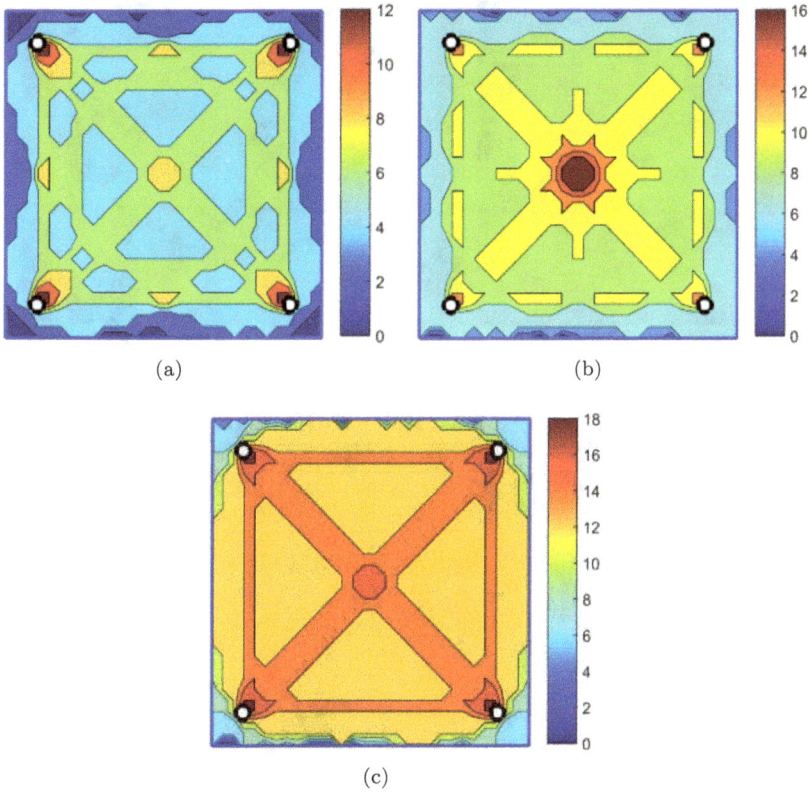

Figure 16.   Influence of the boundary reflection coefficient on intensity of the coverage.[82] (a) $\gamma = 0.75$, (b) $\gamma = 0.4$ and (c) $\gamma = 0.1$.

In addition, the coverage map as well as the intensity of the coverage for the rectangular plate will increase by increasing the transducer numbers from 3 to 5, as shown in Fig. 17.

The input to the proposed optimization is the number of the sensors and the output is the coverage map (including intensity) and the optimal sensor placements. If the intensity of the coverage (related to the PoD of the damage) is not high enough, then the user must increase the number of transducers and run the optimization again until the desired coverage intensity is reached.

The proposed algorithm was tested on a rectangular plate with an opening (representing complex geometry) to find the four optimum sensor locations, which results in a maximum coverage. One hundred possible

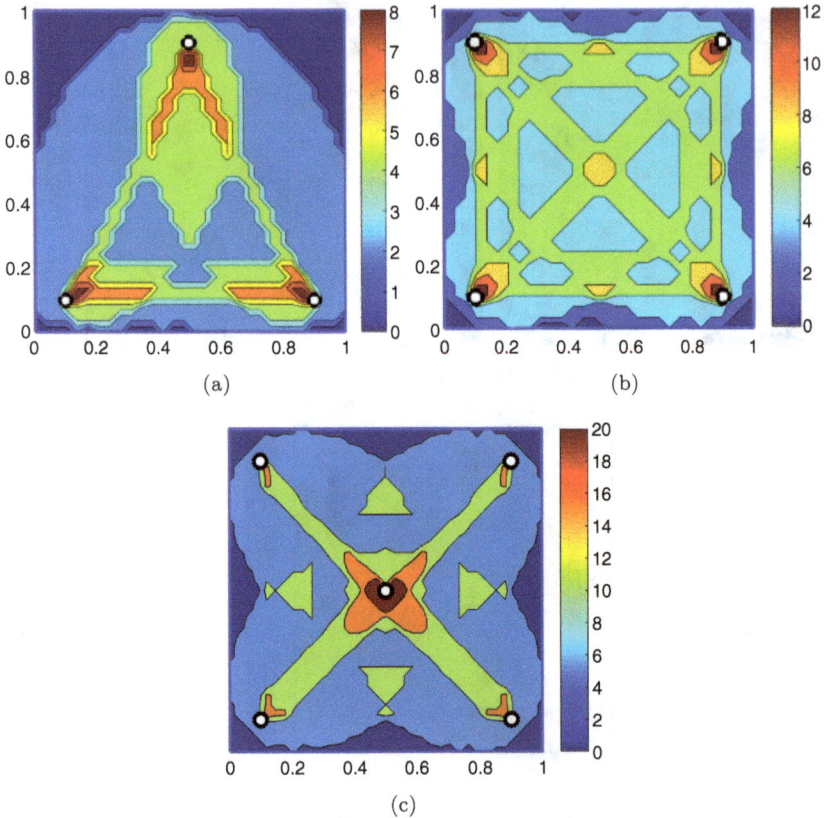

Figure 17.   Coverage area and intensity for different transducer numbers.[82] (a) 3 sensors, (b) 4 sensors and (c) 5 sensors.

sensor locations were assumed in the optimization. The algorithm then searched for the four best transducers that have direct line of sight between each pair. The parameters of the GA were as follows: a population of 30 networks (network of four transducers) for every generation, maximum of 100 generations, mutation rate of 0.02 and crossover fraction on 0.7. The result of the optimization (coverage and the intensity) with the four best sensor locations is shown in Fig. 18. To ensure that this is the optimal configuration, the values of the fitness function was compared to two very similar but sub-optimal configurations. The results showed that the output of the optimization algorithm did result in the maximum coverage area. This confirms that the optimization approach is able to determine an optimal configuration for the sensors, based on the fitness, cross-over

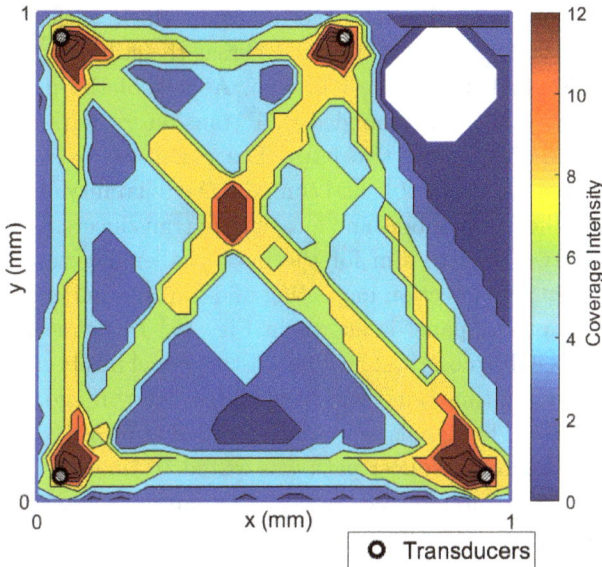

Figure 18. Optimal sensor position for a plate with a hole.

and mutation functions. It can be seen that part of the plate (top right side) is a blind zone, due to the presence of the hole. However, the sensor configuration guarantees a good coverage on the rest of the plate, meaning that damage can be detected in this area. To have a better coverage, the number of sensors should be increased and the new optimal location for the five sensor networks searched.

5. Results and Validation on a Composite Stiffened Panel

This section presents the application of some of the developed methodologies based on UGWs to composite structures.[83] Application of SHM techniques to small coupons have been extensively reported by various authors. Therefore, the focus of this section is to present the results of damage detection on a large composite stiffened panel.

5.1. *Composite Stiffened Panel*

The fuselage panel, which has been tested, is a composite panel with 790×1150 mm dimensions and 1978 mm radius of curvature to the outer surface. The skin and omega hat stiffeners are made of T800/M21 unidirectional pre-preg. The skin is 1.656 mm thick with the following layup: [45/−45/90/0/90/0/90/−45/45], whereas the stiffeners are 1.288 mm thick with [45/−45/0/90/0/−45/45] layup. A part of the panel has been instrumented with 16 surface-bonded PZT transducers, with cyanoacrylate Loctite 401 adhesive, across two bays. The panel was over sensorized to collect as much data as possible and compare the reliability of the detection algorithm for different numbers and locations of transducers. The sensorized area is across two bays to aim for the most challenging case in terms of guided wave damage detection inspecting mid-bay and under the stiffeners. The schematic of the panel is shown in Fig. 19. To define the optimum parameters of the diagnosis system, the pristine panel has been excited with a five cycle Hanning tone-burst with central frequency range of 50–500 kHz in steps of 50 kHz. The signals were then recorded at 60 MS/s at a sampling rate of 0.001 s. Each transducer path was exited 10 times and signals were recorded, bandpass filtered and averaged to ensure repeatability and to minimize the background noise. The range of signals were then used to find the optimum amplitude, attenuation profile and the group velocities of the propagating waves in the stiffened panel.

Once all of the pristine data were recorded, the panel was subjected to 50 J impact under the stiffener (marked by "X" in Fig. 19(b)) from the outer skin to induce BVID, i.e., skin/stringer debonding, which represents the most severe case of un-detected damage for a stiffened panel. In addition, detecting skin/stringer debonding is one of the most challenging damage scenarios for the SHM system to detect due to the propagation profile of the guided waves, such as changes in amplitude and velocity due to the change in the thickness, additional reflections due to stiffener edges and attenuation. A drop tower with a 20-mm radius hemispherical impactor was used.

C-scan for the panel confirmed the presence of BVID as shown in Fig. 20. Both finite element (FE) simulation and experimental results showed that there was a 10% reduction in the compressive strength of the damaged panel.[84]

The signals obtained from the composite stiffened panel (from pristine and damage states) were then used to assess and validate baseline-free and DAS baseline approach in Sections 5.1.2 and 5.1.3.

5.1.1. Instantaneous Baseline Approach

The selected results presented in this section is from the published work of Salmanpour et al.[72] The importance of the baseline-free approach was

(a)

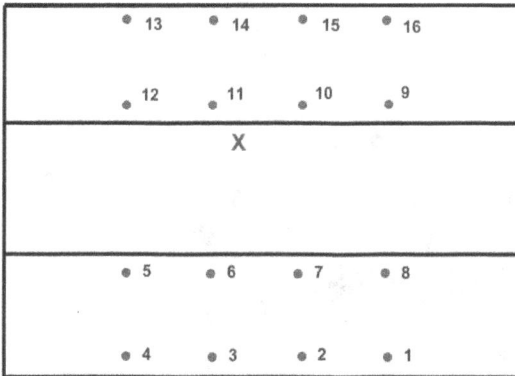

(b)

Figure 19.   (a) Fuselage panel with (b) sensor locations and impact damage.

highlighted in Section 3.2. The novelty of this method is in using an entire area as a baseline to localize damage within an interrogation area. The interrogation area was located at the center of the stiffened panel (area inside transducer network [6–7–10–11] as shown in Fig. 19(b)) while the two available baseline areas of similar geometries are sensor network [5–6–11–12] and network [7–8–9–10]. For this example, the sensor network [5–6–11–12] has been used as baseline for [6–7–10–11] to locate BVID (debonding at the foot of the stiffener runout). The damage was successfully detected using the instantaneous baseline approach as shown by the probability map in Fig. 21.

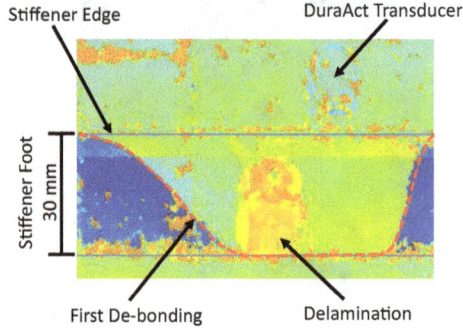

Figure 20.   C-scan of the panel on the location of impact.

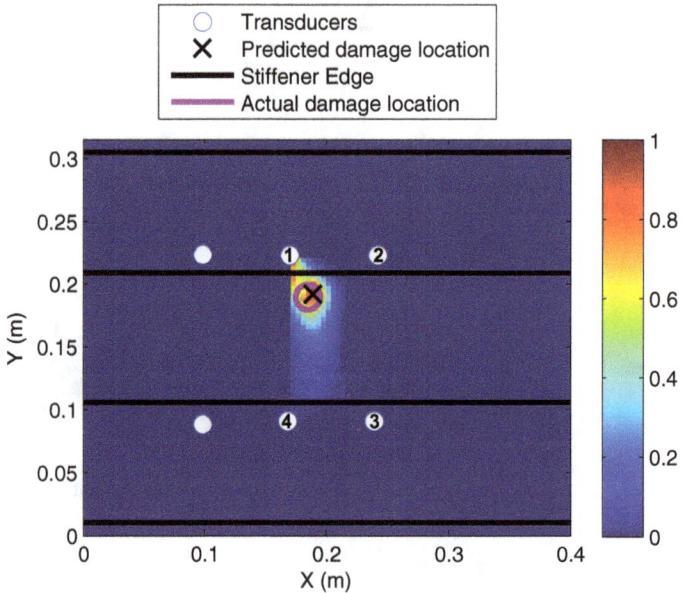

Figure 21.   Damage probability map for the composite stiffened panel.[72]

5.1.2. *DAS Baseline Approach*

For this experiment, each measurement was repeated 10 times and the reversibility of each path was also used to define a noise threshold in the first stage. The DIs for all the paths of the sensor network [1–4–13–16] are presented in Fig. 22 where the noise threshold is indicated by the solid line. The values above the noise threshold indicate changes in the signal

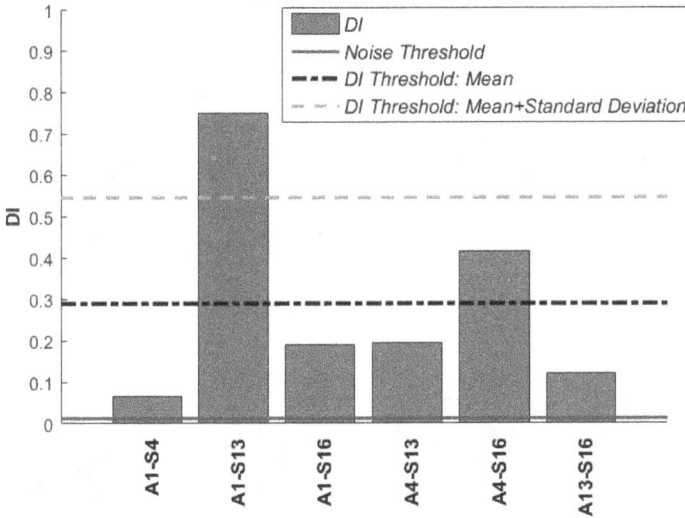

Figure 22. Example of DI threshold value for damage detection.

but not necessarily due to existence of damage. These changes could be attributed to factors such as boundary conditions, loading, temperature, hardware, etc. The second stage is to define a damage threshold to reliably detect the existence of damage. For this reason, the *DI Threshold-Mean* value was identified as the average DI values from all the paths. Any value above this threshold will indicate damage in the close proximity of the path. DI here is represented by root mean square deviation (RMSD) of the residual energy. Since the damage is located on the path 1–13 (see Fig. 19(b)), the DI value for the corresponding path correctly shows the highest value for detecting the damage as well. This can be considered as the first-level damage detection for the structural diagnosis, which results in damage existence and approximate location (path 1–13).

The next step was to assess the accuracy of the DAS detection algorithm for the second level of prognosis, which is damage location. In the first step, all 16 transducers were used and the DI plot for the whole structure was calculated based on Eq. (17) and presented in Fig. 23.

It is clear from the above results that if the intended function of the SHM system is to detect, locate and characterize the damage in addition to its existence, then a dense network of transducers is required. In contrast, for detecting damage existence only, the DI values from four transducers provide values that are significantly above the noise levels. For damage

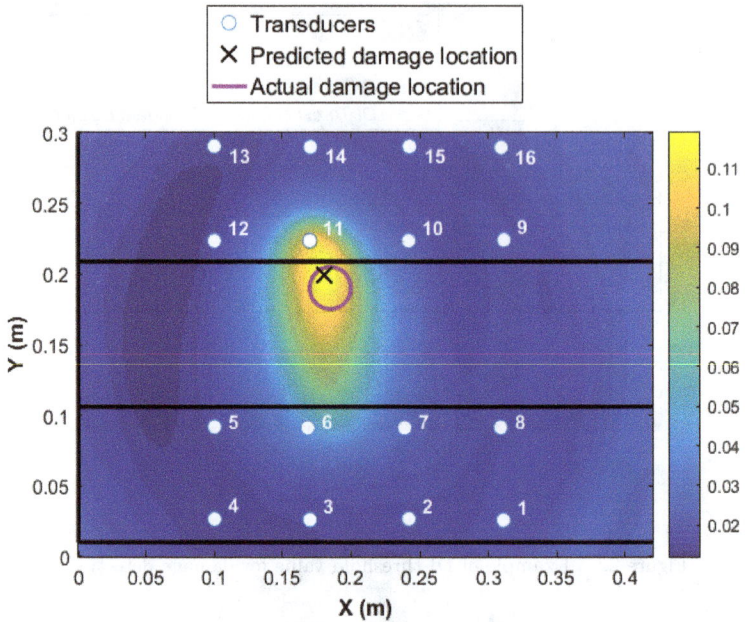

Figure 23.   Damage detection and localization with all 16 transducers following WEAM.[83]

localization to result in accurate damage location, however, not all the transducers are required to be used in the second level of the diagnosis. The fusion algorithm used in Eq. (17) suggests adding the DI contributions for each pixel calculated from each transducer pair. This means that if a transducer pair residual is very low (far from damage), it will not have any positive contributions in the detection algorithm. Therefore, one strategy to increase the intensity of the DI plot and consequently the reliability of the damage detection algorithm is to use only the paths with high DI values in the second level of the diagnosis. To this end a multilevel detection algorithm based on two levels is proposed to increase the reliability of decision-making.[83] The selection criteria for level 2 diagnosis is to use the maximum residual paths (MRPs), which have the maximum effect of damage presence on their paths. The selection of the MRPs is based on EMI measure to identify which transducers are closest to the damage. Based on DAS, the minimum number of the transducers to be used to detect the damage location is three. Therefore, by using the three transducers whose paths go directly or very close to damage, not only the intensity of

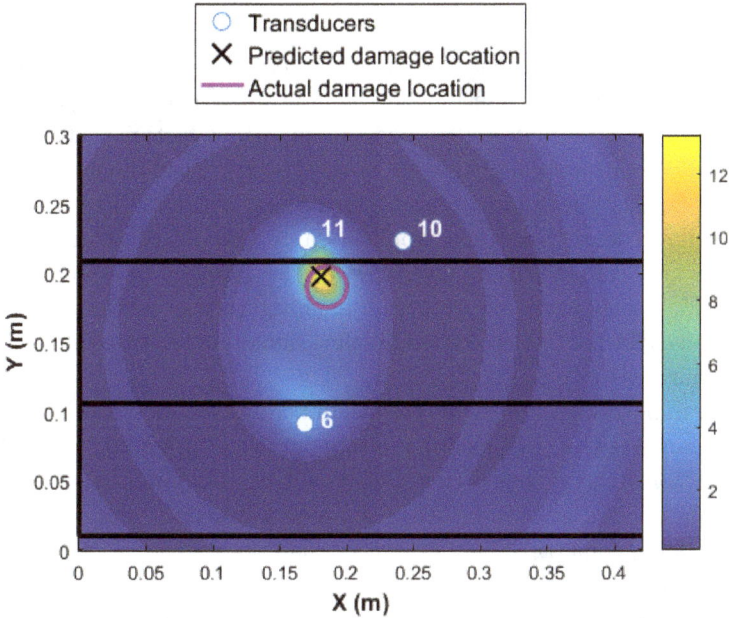

Figure 24. Multilevel damage detection based on MRP transducer network.

the DI could be increased, but also the accuracy of the damage location and sizing was increased as well. Increasing the DI values above the noise threshold is desirable since it increases the reliability of the decision-making and decreases the false alarm rates significantly.

By measuring the residuals for all the paths, it is observed that maximum residual paths are 11–3 and 11–6. This means that the probability of damage being on one of these direct paths is the highest. To select between the transducers 3 and 6, the path lengths 11–6 and 11–3 are compared. It can be seen that the path 11–6 is shorter. Therefore, transducer 6 is chosen as the one closer to the damage site. This means that the damage is inside the transducer network [6–10–11], and the reliability of the damage detection and identification is increased.

The next step of the multilevel diagnosis is to run the WEAM algorithm using the selected transducer network [6–10–11]. The result of the detection is shown in Fig. 24. It can be seen that not only the accuracy of the detection is increased when using the transducer network based on MRP, but also the reliability of detection (increased DI magnitudes) and the accuracy of the damage size have improved significantly.

6. Conclusion

In this chapter, an overview of damage detection algorithms based on UGWs was presented. The fundamentals of guided wave propagation in composite plates were described and the important factors that can affect their propagation were highlighted. Two well-established damage detection algorithms, i.e., the tomography approach and the DAS approach, were explained in detail and their application to composite structures were assessed and compared. An overview of the baseline-free methods was presented, which have the potential to mitigate the environmental and operational effects on the recorded baseline signals an increase the reliability of the diagnosis.

The importance of optimal sensor number and location was highlighted and a summary of the existing strategies were presented. The MAC optimization strategy was then detailed as it has the potential to be easily upscaled to large structures with complex geometries and is based on physical and geometrical factors, rather than obtaining PoD functions for each sensor combination.

At the end, the applications of some of the chosen damage detection algorithms were assessed by testing it on a large-scale composite stiffened panel. It was shown that the multilevel diagnostic approach can significantly increase the reliability of the detection algorithm.

References

1. *Composite Aircraft Structure*, in *AC No: 20-107B* 2009, Federal Aviation Administration: AIR-100.
2. Ewald, V. *et al.*, Design of a structural health monitoring system for a damage tolerance fuselage component. In *7th International Symposium on NDT in Aerospace*, Bremen, 2015.
3. Boller, C., Structural health monitoring — Its association and use. In *New Trends in Structural Health Monitoring*, Springer, 2013, pp. 1–79.
4. Giurgiutiu, V., *Structural Health Monitoring: With Piezoelectric Wafer Active Sensors*, Academic Press, 2007.
5. Takeda, S.-I., Y. Aoki, and Y. Nagao, Damage monitoring of CFRP stiffened panels under compressive load using FBG sensors. *Composite Structures* **94**(3) (2012) 813–819.
6. Sierra-Pérez, J. *et al.*, Damage detection in composite materials structures under variable loads conditions by using fiber Bragg gratings and principal component analysis, involving new unfolding and scaling methods. *Journal of Intelligent Material Systems and Structures* **26**(11) (2015) 1346–1359.

7. Zhao, Y. and F. Ansari, Embedded fiber optic sensor for characterization of interface strains in FRP composite. *Sensors and Actuators A: Physical* **100**(2) (2002) 247–251.

8. Minakuchi, S. *et al.*, Life cycle monitoring of large-scale CFRP VARTM structure by fiber-optic-based distributed sensing. *Composites Part A: Applied Science and Manufacturing* **42**(6) (2011) 669–676.

9. Sánchez, D. M., M. Gresil, and C. Soutis, Distributed internal strain measurement during composite manufacturing using optical fibre sensors. *Composites Science and Technology* **120** (2015) 49–57.

10. Ogisu, T. *et al.*, Development of damage monitoring system for aircraft structure using a PZT actuator/FBG sensor hybrid system. In *Smart Structures and Materials*, International Society for Optics and Photonics, 2004. Proceedings of SPIE Vol. 5388 (SPIE, Bellingham, WA, 2004) 0277-786X/04/$15. doi:10.1117/12/539727.

11. Qing, X. *et al.*, A hybrid piezoelectric/fiber optic diagnostic system for structural health monitoring. *Smart Materials and Structures* **14**(3) (2005) S98.

12. Lin, B. and V. Giurgiutiu, Piezo-optical active sensing with PWAS and FBG sensors for structural health monitoring. In *ASME 2014 Conference on Smart Materials, Adaptive Structures and Intelligent Systems*, American Society of Mechanical Engineers, 2014.

13. Zhong, C., A. Croxford, and P. Wilcox, Remote inspection system for impact damage in large composite structure. In *Proceedings of the Royal Society of America*, The Royal Society, 2015. doi:10.1098/rspa.2014.0631. Published by The Royal Society, Print ISSN 1364-5021; Online ISSN 1471-2946. Published online November 27, 2014. Copyright & Usage © 2014.

14. Flynn, E. B. *et al.*, Maximum-likelihood estimation of damage location in guided-wave structural health monitoring. In *Proceedings of the Royal Society of London A: Mathematical, Physical and Engineering Sciences*, The Royal Society, 2011. doi:10.1098/rspa.2011.0095. Published by The Royal Society, Print ISSN 1364-5021; Online ISSN 1471-2946.

15. Michaels, J. E. and T. E. Michaels, Guided wave signal processing and image fusion for in situ damage localization in plates. *Wave Motion* **44**(6) (2007) 482–492.

16. Kessler, S. S., *Piezoelectric-based in-situ damage detection of composite materials for structural health monitoring systems*, Massachusetts Institute of Technology, Dept. of Aeronautics and Astronautics, 2002.

17. Klepka, A. *et al.*, Impact damage detection in laminated composites by nonlinear vibro-acoustic wave modulations. *Composites Part B: Engineering* **65** (2014) 99–108.

18. Pérez, M. A., L. Gil, and S. Oller, Impact damage identification in composite laminates using vibration testing. *Composite Structures* **108** (2014) 267–276.

19. Thiene, M., M. Zaccariotto, and U. Galvanetto, Application of proper orthogonal decomposition to damage detection in homogeneous plates

and composite beams. *Journal of Engineering Mechanics* **139**(11) (2013) 1539–1550.

20. Giurgiutiu, V. and A. Zagrai, Damage detection in thin plates and aerospace structures with the electro-mechanical impedance method. *Structural Health Monitoring* **4**(2) (2005) 99–118.

21. Wandowski, T., P. Malinowski, and W. Ostachowicz, Delamination detection in CFRP panels using EMI method with temperature compensation. *Composite Structures* (2016).

22. Sharif Khodaei, Z., M. Ghajari, and M. H. Aliabadi, Impact damage detection in composite plates using a self-diagnostic electro-mechanical impedance-based structural health monitoring system. *Journal of Multiscale Modelling* **6**(04) (2015) 1550013.

23. Rose, J. L., *Ultrasonic Waves in Solid Media*, Cambridge University Press, 2004. Avenue of the Americas, New York, NY 10013-2473, USA.

24. Wang, L. and F. Yuan, Group velocity and characteristic wave curves of Lamb waves in composites: Modeling and experiments. *Composites Science and Technology* **67**(7) (2007) 1370–1384.

25. Packo, P., T. Uhl, and W. J. Staszewski, Generalized semi-analytical finite difference method for dispersion curves calculation and numerical dispersion analysis for Lamb waves. *The Journal of the Acoustical Society of America* **136**(3) (2014) 993–1002.

26. Gravenkamp, H. *et al.*, The computation of dispersion relations for three-dimensional elastic waveguides using the scaled boundary finite element method. *Journal of Sound and Vibration* **332**(15) (2013) 3756–3771.

27. Lowe, M. J., Matrix techniques for modeling ultrasonic waves in multilayered media. *IEEE Transactions on Ultrasonics, Ferroelectrics, and Frequency Control* **42**(4) (1995) 525–542.

28. Konstantinidis, G., B. Drinkwater, and P. Wilcox, The temperature stability of guided wave structural health monitoring systems. *Smart Materials and Structures*, **15**(4) (2006) 967.

29. Blaise, E. J. and F.-K. Chang, Built-in damage detection system for sandwich structures under cryogenic temperatures. In *SPIE's 9th Annual International Symposium on Smart Structures and Materials*, International Society for Optics and Photonics, 2002.

30. Konstantinidis, G., P. D. Wilcox, and B. W. Drinkwater, An investigation into the temperature stability of a guided wave structural health monitoring system using permanently attached sensors. *IEEE Sensors Journal* **7**(5) (2007) 905–912.

31. Sabat, R. *et al.*, Temperature dependence of the dielectric, elastic and piezoelectric material coefficients of soft and hard lead zirconate titanate (PZT) ceramics. In *2007 Sixteenth IEEE International Symposium on the Applications of Ferroelectrics*, IEEE, 2007.

32. Hooker, M. W., Properties of PZT-Based piezoelectric Ceramics between-150 and 250 C, 1998. https://ntrs.nasa.gov/search.jsp?R=19980236888.

33. Salmanpour, M. S., Structural health monitoring of composite laminates under operational conditions using piezoelectric transducers. In *Aeronautics*, Imperial College, London, 2017.

34. Lu, Y. and J. E. Michaels, A methodology for structural health monitoring with diffuse ultrasonic waves in the presence of temperature variations. *Ultrasonics* **43**(9) (2005) 717–731.

35. Mallardo, V., M. H. Aliabadi, and Z. S. Khodaei, Optimal sensor positioning for impact localization in smart composite panels. *Journal of Intelligent Material Systems and Structures* **24**(5) (2012) 559–573.

36. Giurgiutiu, V., Structural health monitoring of composite structures with piezoelectric-wafer active sensors. *AIAA* **4**(3) (2011) 565–581.

37. Diamanti, K., C. Soutis, and J. Hodgkinson, Lamb waves for the non-destructive inspection of monolithic and sandwich composite beams. *Composites Part A: Applied Science and Manufacturing* **36**(2) (2005) 189–195.

38. Kessler, S. S., S. M. Spearing, and C. Soutis, Damage detection in composite materials using Lamb wave methods. *Smart Materials and Structures* **11**(2) (2002) 269.

39. Putkis, O., R. Dalton, and A. Croxford, The influence of temperature variations on ultrasonic guided waves in anisotropic CFRP plates. *Ultrasonics* **60** (2015) 109–116.

40. Marzani, A. and S. Salamone, Numerical prediction and experimental verification of temperature effect on plate waves generated and received by piezoceramic sensors. *Mechanical Systems and Signal Processing* **30** (2012) 204–217.

41. Putkis, O. and A. J. Croxford, Continuous baseline growth and monitoring for guided wave SHM. *Smart Materials and Structures* **22**(5) (2013) 055029.

42. Croxford, A. J. *et al.*, Efficient temperature compensation strategies for guided wave structural health monitoring. *Ultrasonics* **50**(4) (2010) 517–528.

43. Clarke, T. and P. Cawley, Enhancing the defect localization capability of a guided wave SHM system applied to a complex structure. *Structural Health Monitoring* **10**(3) (2010) 247–259.

44. Salmanpour, M., Z. S. Khodaei, and M. H. Aliabadi, Guided wave temperature correction methods in structural health monitoring. *Journal of Intelligent Material Systems and Structures* **28**(5) (2016) doi: 10.1177/1045389X16651155.

45. Su, Z. and L. Ye, *Identification of Damage Using Lamb Waves: From Fundamentals to Applications*, Berlin Heidelberg: Springer-Verlag, 2009.

46. Ostachowicz, W. *et al.*, Damage localisation in plate-like structures based on PZT sensors. *Mechanical Systems and Signal Processing* **23**(6) (2009) 1805–1829.

47. Kudela, P., W. Ostachowicz, and A. Żak, Damage detection in composite plates with embedded PZT transducers. *Mechanical Systems and Signal Processing* **22**(6) (2008) 1327–1335.

48. Wang, C. H., J. T. Rose, and F.-K. Chang, A synthetic time-reversal imaging method for structural health monitoring. _Smart Materials and Structures_ **13**(2) (2004) 415.

49. Fromme, P. _et al._, On the development and testing of a guided ultrasonic wave array for structural integrity monitoring. _IEEE Transactions on Ultrasonics, Ferroelectrics, and Frequency Control_ **53**(4) (2006) 777–785.

50. Michaels, J. E., A. J. Croxford, and P. D. Wilcox, Imaging algorithms for locating damage via in situ ultrasonic sensors. In _Sensors Applications Symposium, 2008. SAS 2008_, IEEE, IET, 2008.

51. Sharif Khodaei, Z. and M. H. Aliabadi, Assessment of delay-and-sum algorithms for damage detection in aluminium and composite plates. _Smart Materials and Structures_ **23**(7) (2014) 075007.

52. Lu, G., Y. Li, and G. Song, A delay-and-Boolean-ADD imaging algorithm for damage detection with a small number of piezoceramic transducers. _Smart Materials and Structures_ **25**(9) (2016) 095030.

53. Shan, S. _et al._, Multi-damage localization on large complex structures through an extended delay-and-sum based method. _Structural Health Monitoring_ **15**(1) (2016) 50–64.

54. Hall, J. S., P. Fromme, and J. E. Michaels, Guided wave damage characterization via minimum variance imaging with a distributed array of ultrasonic sensors. _Journal of Nondestructive Evaluation_ **33**(3) (2014) 299–308.

55. Haynes, C. and M. Todd, Enhanced damage localization for complex structures through statistical modeling and sensor fusion. _Mechanical Systems and Signal Processing_ **54** (2015) 195–209.

56. Gao, H. _et al._, Ultrasonic guided wave tomography in structural health monitoring of an aging aircraft wing. In _Proceedings of the ASNT Fall Conference_, 2005.

57. Royer, R. _et al._, Large area corrosion detection in complex aircraft components using Lamb wave tomography. In _Proceedings, of the 6th International Workshop on Structural Health Monitoring_, 2007.

58. Zhao, X. _et al._, Active health monitoring of an aircraft wing with embedded piezoelectric sensor/actuator network: I. Defect detection, localization and growth monitoring. _Smart Materials and Structures_ **16**(4) (2007) 1208.

59. Hettler, J. _et al._, Guided wave tomography based inspection of CFRP plates using a probabilistic reconstruction algorithm. In _Emerging Technologies in Non-Destructive Testing VI: Proceedings of the 6th International Conference on Emerging Technologies in Non-Destructive Testing (Brussels, Belgium, 27–29 May 2015)_, CRC Press, 2015.

60. Hettler, J. _et al._, Linear and nonlinear guided wave imaging of impact damage in CFRP using a probabilistic approach. _Materials_ **9**(11) (2016) 901.

61. Yan, F., R. L. Royer Jr, and J. L. Rose, Ultrasonic guided wave imaging techniques in structural health monitoring. _Journal of Intelligent Material Systems and Structures_ **21**(3) (2010) 377–384.

62. Hall, J. S. and J. E. Michaels, Minimum variance ultrasonic imaging applied to an in situ sparse guided wave array. *IEEE Transactions on Ultrasonics, Ferroelectrics, and Frequency Control* **57**(10) (2010).

63. Qiang, W. and Y. Shenfang, Baseline-free imaging method based on new PZT sensor arrangements. *Journal of Intelligent Material Systems and Structures* **20**(14) (2009) 1663–1673.

64. An, Y.-K. *et al.*, Application of local reference-free damage detection techniques to *in situ* bridges. *Journal of Structural Engineering* **140**(3) (2013) 04013069.

65. Anton, S. R., D. J. Inman, and G. Park, Reference-free damage detection using instantaneous baseline measurements. *AIAA Journal* **47**(8) (2009) 1952–1964.

66. Fink, M. *et al.*, Time-reversed acoustics. *Reports on Progress in Physics* **63**(12) (2000) 1933.

67. Sohn, H. *et al.*, Damage detection in composite plates by using an enhanced time reversal method. *Journal of Aerospace Engineering* **20**(3) (2007) 141–151.

68. Watkins, R. and R. Jha, A modified time reversal method for Lamb wave based diagnostics of composite structures. *Mechanical Systems and Signal Processing* **31** (2012) 345–354.

69. Bruno, C. L. *et al.*, Break of reciprocity principle induced by cracks in concrete: Experimental evidence and applications to nonlinear tomography. In *Proceedings of Meetings on Acoustics XVICNEM*, ASA, 2010.

70. Huang, T. *et al.*, A baseline-free damage detection method based on node displacement mode shape. In *8th European Workshop on Structural Health Monitoring*, Bilbao, Spain, 2016.

71. Park, S., C. Lee, and H. Sohn, Reference-free crack detection using transfer impedances. *Journal of Sound and Vibration* **329**(12) (2010) 2337–2348.

72. Salmanpour, M. S., Z. S. Khodaei, and M. H. Aliabadi, Instantaneous baseline damage localization using sensor mapping. *IEEE Sensors Journal* **17**(2) (2017) 295–301.

73. Croxford, A. J., P. D. Wilcox, and B. W. Drinkwater, Quantification of sensor geometry performance for guided wave SHM. In *SPIE Smart Structures and Materials + Nondestructive Evaluation and Health Monitoring*, International Society for Optics and Photonics, 2009.

74. Malinowski, P., T. Wandowski, and W. Ostachowicz, Damage detection potential of a triangular piezoelectric configuration. *Mechanical Systems and Signal Processing* **25**(7) (2011) 2722–2732.

75. Wandowski, T., P. Malinowski, and W. Ostachowicz, Damage detection with concentrated configurations of piezoelectric transducers. *Smart Materials and Structures* **20**(2) (2011) 025002.

76. Staszewski, W. J. and K. Worden, Overview of optimal sensor location methods for damage detection. In *SPIE's 8th Annual International Symposium on Smart Structures and Materials*, SPIE Defense Security, and Sensing 2001. Vol 4326, pp. 180–187.

77. Guo, H. *et al.*, Optimal placement of sensors for structural health monitoring using improved genetic algorithms. *Smart Materials and Structures* **13**(3) (2004) 528.

78. Flynn, E. and M. Todd, Optimal placement of piezoelectric actuators and sensors for detecting damage in plate structures. *Journal of Intelligent Material Systems and Structures* (2009).

79. Flynn, E. B. and M. D. Todd, A Bayesian approach to optimal sensor placement for structural health monitoring with application to active sensing. *Mechanical Systems and Signal Processing* **24**(4) (2010) 891–903.

80. Salmanpour, M. S., Z. S. Khodaei, and M. H. Aliabadi, Transducer placement optimisation scheme for a delay and sum damage detection algorithm, to appear 2016.

81. Fendzi, C. *et al.*, Optimal sensors placement to enhance damage detection in composite plates. In *7th European Workshop on Structural Health Monitoring*, 2014.

82. Thiene, M., Z. S. Khodaei, and M. H. Aliabadi, Optimal sensor placement for maximum area coverage (MAC) for damage localization in composite structures. *Smart Materials and Structures* **25**(9) (2016) 095037.

83. Sharif Khodaei, Z. and M. H. Aliabadi, A multi-level decision fusion strategy for condition based maintenance of composite structures. *Materials* **9**(9) (2016) 790.

84. Psarras, S. *et al.*, Compression after multiple impacts: Modelling and experimental validation on composite curved stiffened panels, In *Smart Intelligent Aircraft Structures (SARISTU)*, Springer, 2016, pp. 681–689.

Chapter 2

Modeling Guided Wave Propagation in Composite Structures Using Local Interaction Simulation Approach

Yanfeng Shen and Carlos E. S. Cesnik*

*Department of Aerospace Engineering, University of Michigan,
Ann Arbor, 48109, USA*
cesnik@umich.edu

This chapter presents an efficient modeling technique for guided wave propagation in composite structures using a finite difference based numerical scheme — the Local Interaction Simulation Approach (LISA). It starts with an introduction to fundamentals of multimodal dispersive guided waves, the challenges associated with guided wave based Structural Health Monitoring (SHM) techniques in composite structures, and corresponding state-of-the-art modeling strategies. The chapter then describes the LISA formulation derivation from the elastodynamic wave equations considering general anisotropic material properties. Kelvin-Voigt viscoelastic model is integrated to handle the damping effects. The UM-LISA software framework and its implementation with Compute Unified Device Architecture (CUDA) are then discussed. The Absorbing Layers with Increasing Damping (ALID) boundary model is added to the framework to minimize model size and reduce the computational burden in simulations. A case study of ultrasonic guided wave generation and propagation in a highly anisotropic unidirectional composite plate and experimental verification using Scanning Laser Doppler Vibrometry highlight the quality and prowess of the method. Modeling of guided wave propagation in complex-geometry composite panel with stiffeners is presented at the end of the chapter as an example of the applicability of the new technique in practical structural configurations.

1. Introduction

Guided waves are mechanical disturbances propagating in elastic solids when they are guided by the structural boundaries. They arise from the interferential superposition of fundamental pressure wave (P-wave) and shear wave (S-wave) subjected to multiple reflections from the confining surfaces. The structures guiding the wave energy are referred to as waveguides such as plates, rods, rails, or pipes. Guided waves are widely used as interrogating fields in structural health monitoring (SHM) and for non-destructive evaluation (NDE) due to several preferable characteristics: (1) their capability to travel long distances without much energy loss, which makes it possible to interrogate large structural areas from a single location; (2) the sensitivity of though-thickness wave modes to incipient structural changes, which allows the detection of hidden damage within the material; (3) the ability of propagating through structural joints and curvatures, which enables the evaluation of connected complex geometries. However, in practical applications, challenges are found due to the multimodal and dispersive nature of guided waves.

1.1. *Fundamentals of Guided Waves in Isotropic Structures*

To understand the basis of guided waves, consider first an isotropic medium with finite thickness. For these kind of structures, generally four types of guided wave may exist: (1) symmetric Lamb modes; (2) antisymmetric Lamb modes; (3) symmetric shear horizontal (SH) modes; and (4) antisymmetric SH modes.[1] The propagation of Lamb waves is governed by the Rayleigh–Lamb equation:

$$\frac{\tan \eta_S d}{\tan \eta_P d} = \left[\frac{-4\eta_P \eta_S \xi^2}{(\xi^2 - \eta_S^2)^2} \right]^{\pm 1}, \tag{1}$$

where $+1$ exponent corresponds to symmetric Lamb wave modes and -1 exponent corresponds to antisymmetric Lamb wave modes. d is the half thickness, and ξ is the frequency-dependent wavenumber. η_P and η_S are given in Eq. (2). λ and μ are Lame's constants of the material, and ρ is the material density. Moreover

$$\eta_P^2 = \frac{\omega^2}{c_p^2} - \xi^2; \quad \eta_S^2 = \frac{\omega^2}{c_s^2} - \xi^2; \quad c_p = \sqrt{\frac{\lambda + 2\mu}{\rho}}; \quad c_s = \sqrt{\frac{\mu}{\rho}}. \tag{2}$$

The propagation of SH waves has a shear-type particle motion contained in the horizontal plane. According to the coordinate defined in Fig. 1, an SH wave has the particle motion along the y-axis, whereas the wave propagation takes place along the x-axis. The particle motion has only the u_y component. The phase velocity dispersion curve of the SH plate wave

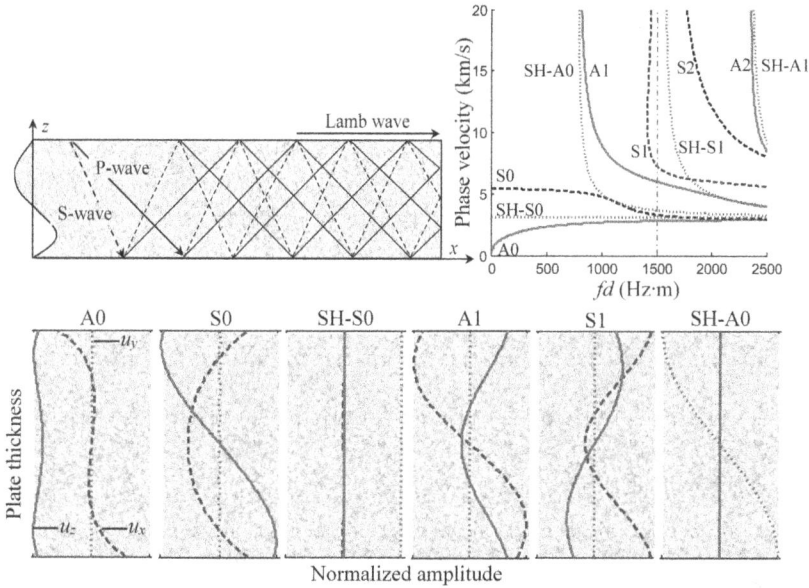

Figure 1. Lamb wave formation from multiple reflections and interferential superposition of fundamental P-wave and S-wave; dispersion curves of guided waves in aluminum plates; representative mode shapes at 1,500 Hz m.

can be calculated as

$$c(\omega) = \frac{c_S}{\sqrt{1 - (\eta d)^2 \left(\frac{c_S}{\omega d}\right)^2}}, \qquad (3)$$

where $\eta^2 = \frac{\omega^2}{c_S^2} - \frac{\omega^2}{c^2}$. By substituting the appropriate eigenvalue, one gets an analytical expression for the wave–speed dispersion curve of each SH wave mode.

Figure 1 shows the Lamb wave formation from multiple reflections and interferential superposition of fundamental P-wave and S-wave, the dispersion curves of guided waves in aluminum plates, as well as the representative guided wave modes at 1,500 Hz m. fd is the frequency half-plate-thickness product. It can be observed that at even low fd values, at least three fundamental wave modes may exist: fundamental symmetric mode (S0), fundamental antisymmetric mode (S0), and fundamental symmetric SH mode (SH–S0). Beyond the cut-off frequency, more and more guided wave modes will appear and may simultaneously participate in the propagation independently. The wave speeds are dispersive, i.e., they change with the corresponding fd value. Figure 2 presents the typical Lamb wave pitch-catch

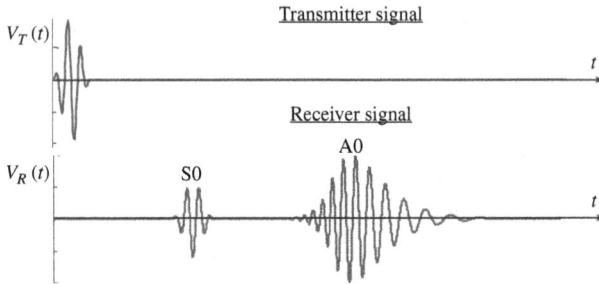

Figure 2.   A typical pitch-catch sensing signal showing multiple elongated wave packets due to multimodal and dispersive characteristics of guided waves.

sensing signal showing two wave packets (S0 and A0) with elongated waveform resulting from the dispersive nature of guided waves, especially obvious for A0 wave mode, which is more dispersive at low fd value.

1.2. Challenges of Guided Wave SHM in Composite Structures

The ever-increasing use of composite components in engineering structures has posed new challenges for implementing effective SHM systems due to the general anisotropic behavior of composite materials. For guided wave-based SHM techniques, these challenges arise in the following aspects:

(1) Guided wave generation shows heavy directional dependence in composite structures. To effectively detect damage at a certain location, sufficient energy of probing waves need to be generated in the interrogation direction. However, the tuning behavior is strongly influenced by the varying material properties along different excitation directions.

(2) Guided waves propagate with different speeds in different directions. This brings considerable challenges for SHM imaging techniques using the time-of-flight (ToF) information and time reversal (TR) method.

(3) Damping is strong, anisotropic and cannot be neglected. Compared with metals, composite materials demonstrate strong and anisotropic damping effects from the fiber/matrix constituents, which attenuates the propagating waves and shrinks their effective interrogation range. This factor will influence SHM techniques based on amplitude/attenuation information.

(4) Complex wave damage interaction scenarios may appear in composites, such as guided wave interaction with matrix cracking, adhesive

deboning, delamination and fiber breaking due to fatigue loading or low velocity impact. These wave damage interaction phenomena need to be well understood to effectively extract the damage information from the sensing signals.

To address these challenges, highly efficient computational methods for ultrasonic guided wave propagation in composite structures are desired for (a) providing guidelines to choose optimal interrogating parameters in SHM system design for the best detection and quantification of damage and (b) unfolding the complicated information of wave damage interaction for effective signal interpretation.

1.3. *Modeling Techniques for Wave Propagation in Composites*

The modeling of guided waves in composite structures, by itself, is a very challenging task. The difficulties arise from the multimodal, dispersive and directional dependent nature of guided waves in composites. The structural details such as layered laminate angles and strong damping effects add to the challenge to accurately depict their influences in a computational model.[2,3] A conventional 3D finite element model containing the detailed information of each lamina may become computationally prohibitive to accommodate the strict accuracy requirements for high frequency and short wavelength ultrasonic waves over long propagation distances.[1,4,5]

Several efficient techniques have been developed for modeling guided waves in laminate composites, such as transfer matrix method (TMM),[6] global matrix method (GMM),[7] and semianalytical finite element (SAFE) method.[8] These techniques provide effective approaches to accurately calculate the dispersion curves of layered structures. However, it is hard to include structural geometric features and damage effects. Ostachowicz *et al.* presented a spectral finite element method (SFEM) for studying guided waves in structures for SHM, achieving considerable higher computational efficiency over the conventional FEM.[9] Li *et al.* used a 3D SFEM to investigate guided wave propagation and interaction with a delamination in a composite plate.[10] However, the heavy computational burden remains if the details of each lamina are included in a SFEM model. Glushkov *et al.* developed a Green's matrix based method to study tuned guided wave excitation and diffraction by obstacles in composite plates.[11]

Delsanto *et al.* proposed a highly efficient numerical technique called the local interaction simulation approach (LISA) using explicit finite difference formulations.[12–14] Lee and Staszewski applied LISA to study wave propagation in isotropic plates and their interaction with damage with a 2D model.[15,16] Sundararaman and Adams extended LISA to 3D problems in unidirectional orthotropic lamina.[17] Packo *et al.* parallelized LISA using compute unified device architecture (CUDA) running on graphical processing unit (GPU), achieving fairly high computational efficiency.[18,19] Nadella and Cesnik extended the 3D LISA formulation to model laminate composites with arbitrary lamination angles.[20] However, the damping effects have been seldom addressed or merely represented using a fairly simple model for isotropic materials in existing LISA literature.[21] To generate guided waves in a LISA model, previous studies generally used prescribed displacements (PDs), which is only a rough estimation of the excitation mechanism. Obenchain *et al.* developed a more accurate approach, using the GMM and analytical solution to calculate the 3D displacement field surrounding a transmitter.[22] This displacement field was then fed into LISA framework through an input boundary. With the pin force excitation assumption, the hybrid GMM/LISA method, however, still cannot capture the local dynamics of the transducers. Nadella and Cesnik integrated piezoelectric coupled field capability in LISA.[23,24] This piezo-coupled LISA can accurately simulate the piezoelectric effects and local dynamics of finite-dimensional transducers.

This chapter presents the LISA modeling technique for the efficient simulation of guided wave generation, propagation and their interaction with damage in composite structures. The basis for UM-LISA formulation is given, followed by addition of damping effects into the LISA equations. The UM-LISA solution procedure and parallel implementation on GPU using CUDA technology are presented. The damping effects in the LISA formulation are used to achieve (1) the accurate modeling of physical anisotropic damping using a Kelvin–Voigt viscoelastic formulation and (2) the construction of absorbing boundary condition (ABC) to effectively shrink the numerical model size. Numerical examples and experimental validation using a scanning laser Doppler vibrometry (SLDV) on carbon fiber composite panels are shown. Finally, the capability of the UM-LISA framework for wave interaction with stiffeners and a delamination damage is presented.

2. UM-LISA Formulation

This section presents the UM-LISA formulation for orthotropic layered composites in non-principal reference frames. The UM-LISA framework is introduced first, followed by the LISA formulation basis for composite lamina with arbitrary orientation. Then, anisotropic damping effects are introduced into the LISA formulation using a Kelvin–Voigt viscoelasticity model.

2.1. *Overview of UM-LISA Framework*

The UM-LISA framework is presented in Fig. 3, showing the LISA derivation procedure as well as the new features. LISA approximates the partial differential elastodynamic wave equations with finite difference quotient expressions. The coefficients in LISA iterative equations (IEs) depend only on the local physical material properties. A sharp interface model (SIM) was used to enforce the stress and displacement continuity between the neighboring cells and nodes. Therefore, changes of material properties in the cells surrounding a computational node can be captured through these coefficients. Guided wave generation is achieved with an efficient frequency domain local FEM. Details of this hybrid approach can be found in Ref. [25]. The 3D displacement field under the transducer was then fed into the global LISA framework as prescribed displacements. In this chapter, damping effects are considered based on the 3D Kelvin–Voigt viscoelasticity model. A viscosity matrix is introduced for a generic lamina with arbitrary stacking angle to capture the directional and coupling damping effects. It should be noted that the new IEs with damping effects require the results from previous three time steps to determine the displacement result at the current time step. A commercial

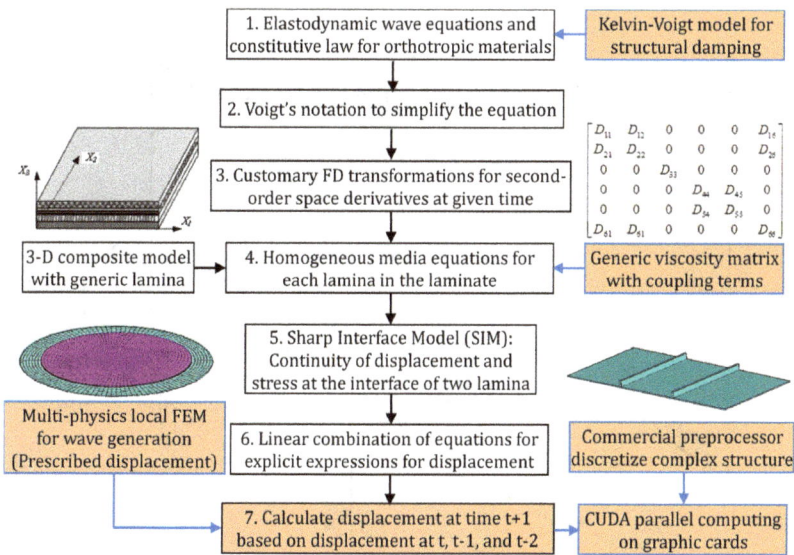

Figure 3.   UM-LISA framework with Kelvin–Voigt viscoelastic damping model. (Colored blocks represent the new implementation aspects to the basic LISA procedure.)

preprocessor from ANSYS was integrated seamlessly into the framework for
computational grid generation and material properties allocation, which
enables to easily model structures with complex geometric features and
material distribution. To ensure high computational efficiency, LISA was
parallelized using CUDA technology and executed on GPU.

2.2. *LISA Formulation Basis for Composite Lamina with Arbitrary Orientation*

The analysis begins with the elastodynamic equations in the displacement
form, which governs the wave motion in elastic solids

$$\partial_l(S_{klmn}w_{m,n}) = \rho\ddot{w}_k \qquad (k, l, m, n = 1, 2, 3), \tag{4}$$

where S is the stiffness tensor, ρ is the material density and w is the
displacement field. The first subscript in the displacement field indicates its
components. Subscripts followed by a comma denote spatial differentiation,
and the dot represents differentiation with respect to time. After the spatial
discretization, a typical computational unit is shown in Fig. 4, showing a
generic node C with 18 nearest neighboring nodes.

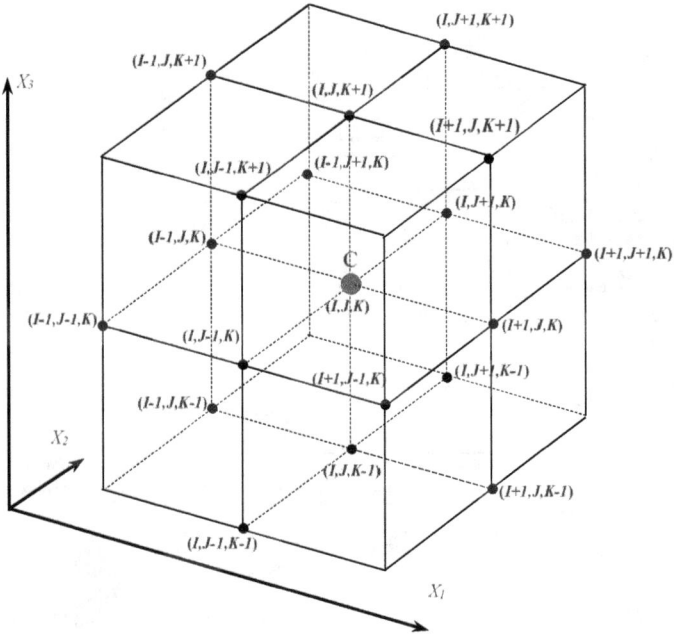

Figure 4. A generic node C at location (I, J, K) with 18 nearest neighboring nodes.

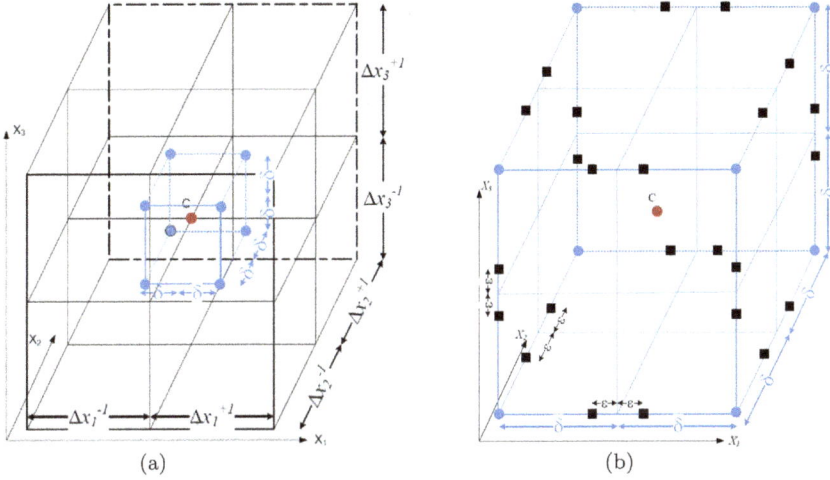

Figure 5. For a generic node C at location (I, J, K), (a) circular blue dots represent the eight points chosen to enforce displacement continuity, and (b) additional points shown as black squares are used to enforce stress continuity conditions.[20]

In addition to the node C and its 18 neighboring nodes, which are physically existing in the computational grid, eight auxiliary points are introduced to assist the derivation at a distance of $\delta \ll \Delta x_i (i = 1, 2, 3)$, given by $(I + \alpha\delta, J + \beta\delta, K + \gamma\delta)$ for $\alpha, \beta, \gamma = \pm 1$, shown in Fig. 5 Voigt's notation is applied to Eq. (4), which combines two indices into a signal index, i.e.,

$$(n, l) \rightarrow \zeta = n\delta_{nl} + (1 - \delta_{nl})(9 - n - l) \quad (\zeta = 1, \ldots, 6)$$
$$(k, m) \rightarrow \eta = k\delta_{km} + (1 - \delta_{km})(9 - k - m) \quad (\eta = 1, \ldots, 6) \tag{5}$$

After applying the Voigt's notation, writing Eq. (4) at the eight auxiliary points for the displacement w_k yields

$$\sum_{m=1}^{3} \sum_{\eta=1}^{6} S_{\eta\xi}^{I+\alpha\delta, J+\beta\delta, K+\gamma\delta} w_{m,\xi}^{I+\alpha\delta, J+\beta\delta, K+\gamma\delta} = \rho \ddot{w}_k^{I+\alpha\delta, J+\beta\delta, K+\gamma\delta}.$$
$$\alpha, \beta, \gamma = \pm 1 \qquad\qquad\qquad k = 1, 2, 3 \tag{6}$$

The resulting equations for a particular component of displacement, w_k ($k = 1, 2, 3$), can be combined at these new points. Continuity of displacement, w_k, is obtained by prescribing displacements at two initial consecutive steps and then enforcing the same acceleration $\ddot{w}_k$ at subsequent time steps.

The second-order derivatives in Eq. (6) are resolved as

$$w_{k,x_1x_1}^{I+\alpha\delta,J+\beta\delta,K+\gamma\delta} = \frac{w_{k,x_1}^{I+\alpha/2,J,K} - w_{k,x_1}^{I+\alpha\delta,J+\beta\delta,K+\gamma\delta}}{\alpha\Delta x_1^\alpha/2},$$

$$w_{k,x_2x_2}^{I+\alpha\delta,J+\beta\delta,K+\gamma\delta} = \frac{w_{k,x_2}^{I,J+\beta/2,K} - w_{k,x_2}^{I+\alpha\delta,J+\beta\delta,K+\gamma\delta}}{\beta\Delta x_2^\beta/2},$$

$$w_{k,x_3x_3}^{I+\alpha\delta,J+\beta\delta,K+\gamma\delta} = \frac{w_{k,x_3}^{I,J,K+\gamma/2} - w_{k,x_3}^{I+\alpha\delta,J+\beta\delta,K+\gamma\delta}}{\gamma\Delta x_3^\gamma/2},$$

$$w_{k,x_1x_2}^{I+\alpha\delta,J+\beta\delta,K+\gamma\delta} = \frac{w_k^{I+\alpha,J+\beta,K} - w_k^{I+\alpha,J,K} - w_k^{I,J+\beta,K} + w_k^{I,J,K}}{\alpha\beta\Delta x_1^\alpha\Delta x_2^\beta},$$

$$w_{k,x_2x_3}^{I+\alpha\delta,J+\beta\delta,K+\gamma\delta} = \frac{w_k^{I,J+\beta,K+\gamma} - w_k^{I,J+\beta,K} - w_k^{I,J,K+\gamma} + w_k^{I,J,K}}{\beta\gamma\Delta x_2^\beta\Delta x_3^\gamma},$$

$$w_{k,x_3x_1}^{I+\alpha\delta,J+\beta\delta,K+\gamma\delta} = \frac{w_k^{I+\alpha,J,K+\gamma} - w_k^{I+\alpha,J,K} - w_k^{I,J,K+\gamma} + w_k^{I,J,K}}{\alpha\gamma\Delta x_1^\alpha\Delta x_3^\gamma},$$

(7)

where the remaining spatial derivatives have similar form. The first-order derivatives in Eq. (7) are substituted with

$$w_{k,x_1}^{I+\alpha/2,J,K} = \frac{w_k^{I+\alpha,J,K} - w_k^{I,J,K}}{\alpha\Delta x_1^\alpha},$$

$$w_{k,x_2}^{I,J+\beta/2,K} = \frac{w_k^{I,J+\beta,K} - w_k^{I,J,K}}{\beta\Delta x_2^\beta},$$

$$w_{k,x_3}^{I,J,K+\gamma/2} = \frac{w_k^{I,J,K+\gamma} - w_k^{I,J,K}}{\gamma\Delta x_3^\gamma},$$

(8)

where the superscript for w_k denotes a particular node in Fig. 4. It should be noted that the first derivative of $w_k^{I+\alpha\delta,J+\beta\delta,K+\gamma\delta}$ for $\alpha,\beta,\gamma = \pm 1$ in Eq. (7) remains unevaluated and is elongated when the stress continuity is enforced. To satisfy the stress continuity condition, additional points are again introduced: $(I + \alpha\varepsilon, J + \beta\delta, K + \gamma\delta)$, $(I + \alpha\delta, J + \beta\varepsilon, K + \gamma\delta)$, and $(I + \alpha\delta, J + \beta\delta, K + \gamma\varepsilon)$, with $\varepsilon \ll \delta \ll \Delta x_i (i = 1, 2, 3)$ represented as black squares in Fig. 5. These additional points do not exist in the mesh grid, nor participate in the computation, but are introduced to accomplish the derivation. The derivatives associated with these points are resolved as

$$w_{k,x_1}^{I+\alpha\varepsilon,J+\beta\delta,K+\gamma\delta} = w_{k,x_1}^{I+\alpha\delta,J+\beta\delta,K+\gamma\delta},$$

$$w_{k,x_2}^{I+\alpha\delta,J+\beta\varepsilon,K+\gamma\delta} = w_{k,x_2}^{I+\alpha\delta,J+\beta\delta,K+\gamma\delta},$$

$$w_{k,x_3}^{I+\alpha\delta,J+\beta\delta,K+\gamma\varepsilon} = w_{k,x_3}^{I+\alpha\delta,J+\beta\delta,K+\gamma\delta},$$

$$w_{k,x_1}^{I+\alpha\delta,J+\beta\varepsilon,K+\gamma\delta} = w_{k,x_1}^{I+\alpha\delta,J+\beta\delta,K+\gamma\varepsilon} = \frac{w_k^{I+\alpha,J,K} - w_k^{I,J,K}}{\alpha\Delta x_1^\alpha}, \qquad (9)$$

$$w_{k,x_2}^{I+\alpha\varepsilon,J+\beta\delta,K+\gamma\delta} = w_{k,x_2}^{I+\alpha\delta,J+\beta\delta,K+\gamma\varepsilon} = \frac{w_k^{I,J+\beta,K} - w_k^{I,J,K}}{\beta\Delta x_2^\beta},$$

$$w_{k,x_3}^{I+\alpha\varepsilon,J+\beta\delta,K+\gamma\delta} = w_{k,x_3}^{I+\alpha\delta,J+\beta\varepsilon,K+\gamma\delta} = \frac{w_k^{I,J,K+\gamma} - w_k^{I,J,K}}{\gamma\Delta x_3^\gamma}.$$

The stress components τ_{pl} can be written as

$$\tau_{pl} = S_{plmn} w_{m,n} \qquad (10)$$

Next, the stress continuity is enforced by matching the stress components at the interfaces of the eight neighboring cells

$$\tau_{p1}^{I+\varepsilon,J+\beta\delta,K+\gamma\delta} = \tau_{p1}^{I-\varepsilon,J+\beta\delta,K+\gamma\delta}$$

$$\tau_{p2}^{I+\alpha\delta,J+\varepsilon,K+\gamma\delta} = \tau_{p2}^{I+\alpha\delta,J-\varepsilon,K+\gamma\delta} \qquad (p = 1, 2, 3). \qquad (11)$$

$$\tau_{p3}^{I+\alpha\delta,J+\beta\delta,K+\varepsilon} = \tau_{p3}^{I+\alpha\delta,J+\beta\delta,K-\varepsilon}$$

Substituting the finite difference quotients in Eq. (6) at the eight surrounding points $(I + \alpha\delta, J + \beta\delta, K + \gamma\delta)$ and using the selected group of stress continuity relations in Eq. (11) to eliminate the unevaluated first-order derivatives, the final LISA equations for the three displacement components are derived for the 3D case as

$$
\begin{aligned}
w_1^{I,J,K,t+1} = {}& -w_1^{I,J,K,t-1} + w_1^{I,J,k} \\[6pt]
& -\frac{2\chi}{8} w_1^{I,J,K} \sum_{\alpha,\beta,\gamma=\pm 1} \left[\eta_x^2 \tilde{S}_{11} + \eta_y^2 \tilde{S}_{66} + \eta_z^2 \tilde{S}_{55} \right] \\[6pt]
& +\frac{\chi}{8} \sum_{\alpha,\beta,\gamma=\pm 1} \left[2\eta_x^2 \tilde{S}_{11} w_1^{I+\alpha,J,K} + 2\eta_y^2 \tilde{S}_{66} w_1^{I,J+\beta,K} \right. \\[6pt]
& \left. + 2\eta_z^2 \tilde{S}_{55} w_1^{I,J,K+\gamma} \right] \\[6pt]
& +\frac{\chi}{8} \sum_{\alpha,\beta,\gamma=\pm 1} \left[\alpha\beta\eta_x\eta_y \left(\tilde{S}_{12} + \tilde{S}_{66} \right) \left(w_2^{I+\alpha,J+\beta,K} - w_2^{I,J,K} \right) \right] \\[6pt]
& +\frac{\chi}{8} \sum_{\alpha,\beta,\gamma=\pm 1} \left[\alpha\beta\eta_x\eta_y \left(\tilde{S}_{12} - \tilde{S}_{66} \right) \left(w_2^{I,J+\beta,K} - w_2^{I+\alpha,J,K} \right) \right] \\[6pt]
& +\frac{\chi}{8} \sum_{\alpha,\beta,\gamma=\pm 1} \left[\alpha\gamma\eta_x\eta_z \left(\tilde{S}_{13} + \tilde{S}_{55} \right) \left(w_3^{I+\alpha,J,K+\gamma} - w_3^{I,J,K} \right) \right] \\[6pt]
& +\frac{\chi}{8} \sum_{\alpha,\beta,\gamma=\pm 1} \left[\alpha\gamma\eta_x\eta_z \left(\tilde{S}_{13} - \tilde{S}_{55} \right) \left(w_3^{I,J,k} - w_3^{I+\alpha,J,K} \right) \right] \\[6pt]
& -\frac{2\chi}{8} \sum_{\alpha,\beta,\gamma=\pm 1} \left[\alpha\beta\eta_x\eta_y \tilde{S}_{16} \left(w_1^{I,J,K} - w_1^{I+\alpha,J+\beta,K} \right) \right] \\[6pt]
& -\frac{2\chi}{8} w_2^{I,J,K} \sum_{\alpha,\beta,\gamma=\pm 1} \left[\eta_x^2 \tilde{S}_{16} + \eta_y^2 \tilde{S}_{26} \right] \\[6pt]
& -\frac{\chi}{8} \sum_{\alpha,\beta,\gamma=\pm 1} \left[\beta\gamma\eta_y\eta_z \left(\tilde{S}_{36} + \tilde{S}_{45} \right) w_3^{I,J,K} \right] \\[6pt]
& +\frac{2\chi}{8} \sum_{\alpha,\beta,\gamma=\pm 1} \left[\eta_x^2 \tilde{S}_{16} w_2^{I+\alpha,J,K} + \eta_y^2 \tilde{S}_{26} w_2^{I,J+\beta,K} \right] \\[6pt]
& +\frac{\chi}{8} \sum_{\alpha,\beta,\gamma=\pm 1} \left[\beta\gamma\eta_y\eta_z \tilde{S}_{36} \right. \\[6pt]
& \left. \times \left(w_3^{I,J+\beta,K+\gamma} + w_3^{I,J,K+\gamma} - w_3^{I,J+\beta,K} \right) \right]
\end{aligned}
$$

$$+ \frac{\chi}{8} \sum_{\alpha,\beta,\gamma=\pm 1} \left[\beta\gamma\eta_y\eta_z \tilde{S}_{45} \right.$$

$$\times \left(w_3^{I,J+\beta,K+\gamma} - w_3^{I,J,K+\gamma} + w_3^{I,J+\beta,K} \right) \right]$$

$$+ \frac{2\chi}{8} \sum_{\alpha,\beta,\gamma=\pm 1} \left[\eta_z^2 \tilde{S}_{45} \left(w_2^{I,J,K+\gamma} - w_2^{I,J,K} \right) \right], \tag{12}$$

$$w_2^{I,J,K,t+1} = - w_2^{I,J,K,t-1} + w_2^{I,J,K}$$

$$- \frac{2\chi}{8} w_2^{I,J,K} \sum_{\alpha,\beta,\gamma=\pm 1} \left[\eta_x^2 \tilde{S}_{66} + \eta_y^2 \tilde{S}_{22} + \eta_z^2 \tilde{S}_{44} \right]$$

$$+ \frac{\chi}{8} \sum_{\alpha,\beta,\gamma=\pm 1} \left[2\eta_x^2 \tilde{S}_{66} w_2^{I+\alpha,J,K} + 2\eta_y^2 \tilde{S}_{22} w_2^{I,J+\beta,K} \right.$$

$$\left. + 2\eta_z^2 \tilde{S}_{44} w_2^{I,J,K+\gamma} \right]$$

$$+ \frac{\chi}{8} \sum_{\alpha,\beta,\gamma=\pm 1} \left[\alpha\beta\eta_x\eta_y \left(\tilde{S}_{12} + \tilde{S}_{66} \right) \left(w_1^{I+\alpha,J+\beta,K} - w_1^{I,J,K} \right) \right]$$

$$+ \frac{\chi}{8} \sum_{\alpha,\beta,\gamma=\pm 1} \left[\alpha\beta\eta_x\eta_y \left(\tilde{S}_{12} - \tilde{S}_{66} \right) \left(w_1^{I+\alpha,J,K} - w_1^{I,J+\beta,K} \right) \right]$$

$$+ \frac{\chi}{8} \sum_{\alpha,\beta,\gamma=\pm 1} \left[\beta\gamma\eta_y\eta_z \left(\tilde{S}_{23} + \tilde{S}_{44} \right) \left(w_3^{I,J+\beta,K+\gamma} - w_3^{I,J,K} \right) \right]$$

$$+ \frac{\chi}{8} \sum_{\alpha,\beta,\gamma=\pm 1} \left[\beta\gamma\eta_y\eta_z \left(\tilde{S}_{23} - \tilde{S}_{44} \right) \left(w_3^{I,J,K+\gamma} - w_3^{I,J+\beta,K} \right) \right]$$

$$- \frac{2\chi}{8} \sum_{\alpha,\beta,\gamma=\pm 1} \left[\alpha\beta\eta_x\eta_y \tilde{S}_{26} \left(w_2^{I,J,K} - w_2^{I+\alpha,J+\beta,K} \right) \right]$$

$$- \frac{2\chi}{8} w_1^{I,J,K} \sum_{\alpha,\beta,\gamma=\pm 1} \left[\eta_x^2 \tilde{S}_{16} + \eta_y^2 \tilde{S}_{26} \right]$$

$$- \frac{\chi}{8} w_1^{I,J,K} \sum_{\alpha,\beta,\gamma=\pm 1} \left[\alpha\gamma\eta_x\eta_z \left(\tilde{S}_{36} + \tilde{S}_{45} \right) w_3^{I,J,K} \right]$$

$$+ \frac{2\chi}{8} \sum_{\alpha,\beta,\gamma=\pm 1} \left[\eta_x^2 \tilde{S}_{16} w_1^{I+\alpha,J,K} + \eta_y^2 \tilde{S}_{26} w_1^{I,J+\beta,K} \right]$$

$$+ \frac{\chi}{8} \sum_{\alpha,\beta,\gamma=\pm 1} \left[\alpha\gamma\eta_x\eta_z \tilde{S}_{36} \right.$$

$$\times \left(w_3^{I+\alpha,J,K+\gamma} + w_3^{I,J,K+\gamma} - w_3^{I+\alpha,J,K} \right) \right]$$

$$+ \frac{\chi}{8} \sum_{\alpha,\beta,\gamma=\pm 1} \left[\alpha\gamma\eta_x\eta_z \tilde{S}_{45} \right.$$

$$\times \left(w_3^{I+\alpha,J,K+\gamma} - w_3^{I,J,K+\gamma} + w_3^{I+\alpha,J,K} \right) \right]$$

$$+ \frac{2\chi}{8} \sum_{\alpha,\beta,\gamma=\pm 1} \left[\eta_z^2 \tilde{S}_{45} \left(w_1^{I,J,K+\gamma} - w_1^{I,J,K} \right) \right], \tag{13}$$

$$w_3^{I,J,K,t+1} = - w_3^{I,J,K,t-1} + w_3^{I,J,K}$$

$$- \frac{2\chi}{8} w_3^{I,J,K} \sum_{\alpha,\beta,\gamma=\pm 1} \left[\eta_x^2 \tilde{S}_{55} + \eta_y^2 \tilde{S}_{44} + \eta_z^2 \tilde{S}_{33} \right]$$

$$+ \frac{\chi}{8} \sum_{\alpha,\beta,\gamma=\pm 1} \left[2\eta_x^2 \tilde{S}_{55} w_3^{I+\alpha,J,K} + 2\eta_y^2 \tilde{S}_{44} w_3^{I,J+\beta,K} \right.$$

$$+ 2\eta_z^2 \tilde{S}_{33} w_3^{I,J,K+\gamma} \right]$$

$$+ \frac{\chi}{8} \sum_{\alpha,\beta,\gamma=\pm 1} \left[\beta\gamma\eta_y\eta_z \left(\tilde{S}_{23} + \tilde{S}_{44} \right) \left(w_2^{I,J+\beta,K+\gamma} - w_2^{I,J,K} \right) \right]$$

$$+ \frac{\chi}{8} \sum_{\alpha,\beta,\gamma=\pm 1} \left[\beta\gamma\eta_y\eta_z \left(\tilde{S}_{23} - \tilde{S}_{44} \right) \left(w_2^{I,J+\beta,K} - w_2^{I,J,K+\gamma} \right) \right]$$

$$+ \frac{\chi}{8} \sum_{\alpha,\beta,\gamma=\pm 1} \left[\alpha\gamma\eta_x\eta_z \left(\tilde{S}_{13} + \tilde{S}_{55} \right) \left(w_1^{I+\alpha,J,K+\gamma} - w_1^{I,J,K} \right) \right]$$

$$+ \frac{\chi}{8} \sum_{\alpha,\beta,\gamma=\pm 1} \left[\alpha\gamma\eta_x\eta_z \left(\tilde{S}_{13} - \tilde{S}_{55} \right) \left(w_1^{I+\alpha,J,K} - w_1^{I,J,K+\gamma} \right) \right]$$

$$-\frac{\chi}{8} \sum_{\alpha,\beta,\gamma=\pm 1} \left[\beta\gamma\eta_y\eta_z \left(\tilde{S}_{36} + \tilde{S}_{45} \right) \left(w_1^{I,J,K} - w_1^{I,J+\beta,K+\gamma} \right) \right]$$

$$-\frac{\chi}{8} \sum_{\alpha,\beta,\gamma=\pm 1} \left[\beta\gamma\eta_y\eta_z \left(\tilde{S}_{36} - \tilde{S}_{45} \right) \left(w_1^{I,J,K+\gamma} - w_1^{i,J+\beta,K} \right) \right]$$

$$-\frac{\chi}{8} \sum_{\alpha,\beta,\gamma=\pm 1} \left[\alpha\gamma\eta_x\eta_z \left(\tilde{S}_{36} + \tilde{S}_{45} \right) \left(w_2^{I,J,K} - w_2^{I+\alpha,J,K+\gamma} \right) \right]$$

$$-\frac{\chi}{8} \sum_{\alpha,\beta,\gamma=\pm 1} \left[\alpha\gamma\eta_x\eta_z \left(\tilde{S}_{36} - \tilde{S}_{45} \right) \left(w_2^{I,J,K+\gamma} - w_2^{I+\alpha,J,K} \right) \right]$$

$$+\frac{2\chi}{8} \sum_{\alpha,\beta,\gamma=\pm 1} \left[\alpha\beta\eta_x\eta_y\tilde{S}_{45} \left(w_3^{I+\alpha,J+\beta,K} - w_3^{I,J,K} \right) \right], \qquad (14)$$

where $\eta_x = 1/\Delta x_1^\alpha$, $\eta_y = 1/\Delta x_2^\beta$, $\eta_z = 1/\Delta x_3^\gamma$, and Δx_1^α, Δx_2^β, Δx_3^γ are the spatial steps along X_1, X_2 and X_3 axes. The current time t is assumed where it is not mentioned. $\tilde{S}_{11} = S_{11}(I + \alpha, J + \beta, K + \gamma)$ represents the stiffness tensor component S_{11} in one of the eight cells surrounding the target node C depending on the choice of (α, β, γ) from $(+1, -1)$, and similar expressions hold for the other damping terms. $\chi = (\Delta t^2/\bar{\rho})$, where Δt is the time step used in the simulation and $\bar{\rho}$ is the average density of all the eight cells surrounding point C as shown in Fig. 4.

2.3. *Adding Damping Effects into the LISA Formulation*

The LISA formulation can be further extended to include damping effects for composite structures based on 3D Kelvin–Voigt viscoelasticity theory. Therefore, the damping effects are not directly related to nodal velocities as presented in Ref. [21]. Instead, they are proportional to the strain rates $\dot{\varepsilon}_{mn}$. For viscoelastic materials, the stress components σ_{kl} are the linear combination of elastic and viscous components and can be written as

$$\sigma_{kl} = S_{klmn}\varepsilon_{mn} + D_{klmn}\dot{\varepsilon}_{mn}, \qquad (15)$$

where S_{klmn} is the stiffness matrix and D_{klmn} is the viscosity matrix. Substituting Eq. (15) into the elastodynamic equations yields the formulation of wave propagation in viscoelastic materials

$$\partial_l \left(S_{klmn}w_{m,n} + D_{klmn}\dot{w}_{m,n} \right) = \rho\ddot{w}_k, \quad k,l,m,n = 1,2,3. \qquad (16)$$

This damping formulation is capable of modeling the anisotropic damping behavior of a lamina with an arbitrary stacking angle, i.e., it can capture the fact that the damping properties differ along various propagation directions and the damping effects may couple between different directions in a lamina with an arbitrary stacking angle. Saravanos and Chamis investigated the anisotropic damping effects of unidirectional fiber-reinforced composites using a unified micromechanics theory.[26] It was found that when the lamina orientation coincides with the global computational coordinates or the loading axis, the viscosity matrix contains only diagonal terms, and the damping effects in different directions are decoupled. When the lamina is placed at an arbitrary angle, coupling terms will appear. The transformation from on-axis viscosity matrix into off-axis viscosity matrix.

$$
\begin{bmatrix}
D_{11}^o & 0 & 0 & 0 & 0 & 0 \\
0 & D_{22}^o & 0 & 0 & 0 & 0 \\
0 & 0 & D_{33}^o & 0 & 0 & 0 \\
0 & 0 & 0 & D_{44}^o & 0 & 0 \\
0 & 0 & 0 & 0 & D_{55}^o & 0 \\
0 & 0 & 0 & 0 & 0 & D_{66}^o
\end{bmatrix}
\xrightarrow[\text{transform}]{D = \mathbf{T}D^o\mathbf{T}^{-1}}
\begin{bmatrix}
D_{11} & D_{12} & 0 & 0 & 0 & D_{16} \\
D_{21} & D_{22} & 0 & 0 & 0 & D_{26} \\
0 & 0 & D_{33} & 0 & 0 & 0 \\
0 & 0 & 0 & D_{44} & D_{45} & 0 \\
0 & 0 & 0 & D_{54} & D_{55} & 0 \\
D_{61} & D_{61} & 0 & 0 & 0 & D_{66}
\end{bmatrix}
$$

$$(17)$$

where T is the transformation matrix for second-order tensors. In general, for on-axis loading condition, $D_{11}^o \neq D_{22}^o = D_{33}^o$ (related to dilatational/extensional damping) and $D_{44}^o \neq D_{55}^o = D_{66}^o$ (related to distortional/shear damping). For isotropic materials, $D_{11}^o = D_{22}^o = D_{33}^o$ and $D_{44}^o = D_{55}^o = D_{66}^o$. It should be noted that, after the transformation, the viscosity matrix stays symmetric, i.e., $D_{kl} = D_{lk}$. Thus, four additional coupling coefficients D_{12}, D_{16}, D_{26} and D_{45} should be considered in the final formulation.

After a careful observation of Eq. (16), one might notice that the addition of the viscous terms does not alter the deriving procedure of the LISA equations shown in Section 2 for the following two reasons: (1) the elastic stresses $S_{klmn}w_{m,n}$ and the viscous stresses $D_{klmn}\dot{w}_{m,n}$ share the same finite difference schemes in the space domain; (2) the space domain and time domain finite difference procedures are independent. Thus, the overall derivation procedure for the second-order space derivatives follows Section 2 and is omitted here. The main difference arises when applying the sharp interface model (SIM) to impose stress continuity. The total viscoelastic stress in Eq. (15) should be matched, instead of merely the elastic part in Eq. (10). Up to this point, the space derivatives are replaced by corresponding customary finite difference transformations, leaving only time derivatives. Then, the second-order time derivatives, associated with $\rho\ddot{w}_k$, are replaced by the conventional second-order central difference expression, while the first-order derivatives resulted from the viscous terms are approximated by the second-order left-sided finite difference expression. Such a transformation for the first-order time derivatives maintains the LISA formulation in an explicit format and keeps a second-order time marching accuracy. The implementation of this left-sided expression requires the results from previous three time steps to determine the displacement result at current time step, instead of previous two time steps in the conventional LISA IEs. The extended LISA equations with damping effects are written in the form of the superposition of contributions from stiffness (elasticity) and damping (viscosity), i.e.,

$$w_1^{I,J,K,t+1} = w_1^{I,J,K,t+1}(S) + w_1^{I,J,K,t+1}(D), \tag{18}$$

$$w_2^{I,J,K,t+1} = w_2^{I,J,K,t+1}(S) + w_2^{I,J,K,t+1}(D), \tag{19}$$

$$w_3^{I,J,K,t+1} = w_3^{I,J,K,t+1}(S) + w_3^{I,J,K,t+1}(D), \tag{20}$$

where the first three superscripts denote the node identity. The fourth superscript represents the time step. The subscript represents the coordinate direction. For instance, $w_1^{I,J,K,t+1}$ is the displacement of node (I, J, K) at time $t + 1$ in X_1 direction. The S and D between parenthesis designate the contribution from stiffness and damping, respectively. For the stiffness contribution, $w^{I,J,K,t+1}(S)$, the readers are referred to Ref. [20], which is the foundation of the 3D LISA formulation for composite structures. The extended damping contributions, $w_m^{I,J,K,t+1}(D)$, for three coordinate directions are given by

$$w_1^{I,J,K,t+1}(D) = -\frac{2\chi}{8}\dot{w}_1^{I,J,K}\sum_{\alpha,\beta,\gamma=\pm1}\left[\eta_x^2\tilde{D}_{11} + \eta_y^2\tilde{D}_{66} + \eta_z^2\tilde{D}_{55}\right]$$

$$+ \frac{\chi}{8}\sum_{\alpha,\beta,\gamma=\pm1}\left[2\eta_x^2\tilde{D}_{11}\dot{w}_1^{I+\alpha,J,K} + 2\eta_y^2\tilde{D}_{66}\dot{w}_1^{I,J+\beta,K}\right.$$

$$+ 2\eta_z^2\tilde{D}_{55}\dot{w}_1^{I,J,K+\gamma}\right]$$

$$+ \frac{\chi}{8}\sum_{\alpha,\beta,\gamma=\pm1}\left[\alpha\beta\eta_x\eta_y\left(\tilde{D}_{12} + \tilde{D}_{66}\right)\right.$$

$$\times \left(\dot{w}_2^{I+\alpha,J+\beta,K} - \dot{w}_2^{I,J,K} \right) \Big]$$

$$+ \frac{\chi}{8} \sum_{\alpha,\beta,\gamma=\pm 1} \Big[\alpha\beta\eta_x\eta_y \left(\tilde{D}_{12} - \tilde{D}_{66} \right)$$

$$\times \left(\dot{w}_2^{I,J+\beta,K} - \dot{w}_2^{I+\alpha,J,K} \right) \Big]$$

$$+ \frac{\chi}{8} \sum_{\alpha,\beta,\gamma=\pm 1} \Big[\alpha\gamma\eta_x\eta_z \tilde{D}_{55}$$

$$\times \left(\dot{w}_3^{I+\alpha,J,K+\gamma} - \dot{w}_3^{I,J,K} - \dot{w}_3^{I,J,K+\gamma} + \dot{w}_3^{I+\alpha,J,K} \right) \Big]$$

$$- \frac{2\chi}{8} \sum_{\alpha,\beta,\gamma=\pm 1} \Big[\alpha\beta\eta_x\eta_y \tilde{D}_{16} \left(\dot{w}_1^{I,J,K} - \dot{w}_1^{I+\alpha,J+\beta,K} \right) \Big]$$

$$- \frac{2\chi}{8} \dot{w}_2^{I,J,K} \sum_{\alpha,\beta,\gamma=\pm 1} \Big[\eta_x^2 \tilde{D}_{16} + \eta_y^2 \tilde{D}_{26} \Big]$$

$$- \frac{\chi}{8} \dot{w}_3^{I,J,K} \sum_{\alpha,\beta,\gamma=\pm 1} \Big[\beta\gamma\eta_y\eta_z \tilde{D}_{45} \Big]$$

$$+ \frac{2\chi}{8} \sum_{\alpha,\beta,\gamma=\pm 1} \Big[\eta_x^2 \tilde{D}_{16} \dot{w}_2^{I+\alpha,J,K} + \eta_y^2 \tilde{D}_{26} \dot{w}_2^{I,J+\beta,K} \Big]$$

$$+ \frac{\chi}{8} \sum_{\alpha,\beta,\gamma=\pm 1} \Big[\beta\gamma\eta_y\eta_z \tilde{D}_{45}$$
$$\times \left(\dot{w}_3^{I,J+\beta,K+\gamma} - \dot{w}_3^{I,J,K+\gamma} + \dot{w}_3^{I,J+\beta,K} \right) \Big]$$

$$+ \frac{2\chi}{8} \sum_{\alpha,\beta,\gamma=\pm 1} \Big[\eta_z^2 \tilde{D}_{45} \left(\dot{w}_2^{I,J,K+\gamma} - \dot{w}_2^{I,J,K} \right) \Big], \tag{21}$$

$$w_2^{I,J,K,t+1}(D) = - \frac{2\chi}{8} \dot{w}_2^{I,J,K} \sum_{\alpha,\beta,\gamma=\pm 1} \Big[\eta_x^2 \tilde{D}_{66} + \eta_y^2 \tilde{D}_{22} + \eta_z^2 \tilde{D}_{44} \Big]$$

$$+ \frac{\chi}{8} \sum_{\alpha,\beta,\gamma=\pm 1} \Big[2\eta_x^2 \tilde{D}_{66} \dot{w}_2^{I+\alpha,J,K} + 2\eta_y^2 \tilde{D}_{22} \dot{w}_2^{I,J+\beta,K}$$

$$+ 2\eta_z^2 \tilde{D}_{44} \dot{w}_2^{I,J,K+\gamma} \Big]$$

$$+ \frac{\chi}{8} \sum_{\alpha,\beta,\gamma=\pm 1} \Big[\alpha\beta\eta_x\eta_y \left(\tilde{D}_{12} + \tilde{D}_{66} \right)$$

$$\times \left(\dot{w}_1^{I+\alpha,J+\beta,K} - \dot{w}_1^{I,J,K} \right) \Big]$$

$$+ \frac{\chi}{8} \sum_{\alpha,\beta,\gamma=\pm 1} \Big[\alpha\beta\eta_x\eta_y \left(\tilde{D}_{12} - \tilde{D}_{66} \right)$$

$$\times \left(\dot{w}_1^{I+\alpha,J,K} - \dot{w}_1^{I,J+\beta,K} \right) \Big]$$

$$+ \frac{\chi}{8} \sum_{\alpha,\beta,\gamma=\pm 1} \Big[\beta\gamma\eta_y\eta_z \tilde{D}_{44}$$

$$\times \left(\dot{w}_3^{I,J+\beta,K+\gamma} - \dot{w}_3^{I,J,K} - \dot{w}_3^{I,J,K+\gamma} + \dot{w}_3^{I,J+\beta,K} \right) \Big]$$

$$- \frac{2\chi}{8} \sum_{\alpha,\beta,\gamma=\pm 1} \Big[\alpha\beta\eta_x\eta_y \tilde{D}_{26} \left(\dot{w}_2^{I,J,K} - \dot{w}_2^{I+\alpha,J+\beta,K} \right) \Big]$$

$$- \frac{2\chi}{8} \dot{w}_1^{I,J,K} \sum_{\alpha,\beta,\gamma=\pm 1} \Big[\eta_x^2 \tilde{D}_{16} + \eta_y^2 \tilde{D}_{26} \Big]$$

$$- \frac{\chi}{8} \dot{w}_3^{I,J,K} \sum_{\alpha,\beta,\gamma=\pm 1} \Big[\alpha\gamma\eta_x\eta_z \tilde{D}_{45} \Big]$$

$$+ \frac{2\chi}{8} \sum_{\alpha,\beta,\gamma=\pm 1} \Big[\eta_x^2 \tilde{D}_{16} \dot{w}_1^{I+\alpha,J,K} + \eta_y^2 \tilde{D}_{26} \dot{w}_1^{I,J+\beta,K} \Big]$$

$$+ \frac{\chi}{8} \sum_{\alpha,\beta,\gamma=\pm 1} \Big[\alpha\gamma\eta_x\eta_z \tilde{D}_{45}$$

$$\times \left(\dot{w}_3^{I+\alpha,J,K+\gamma} - \dot{w}_3^{I,J,K+\gamma} + \dot{w}_3^{I+\alpha,J,K} \right) \Big]$$

$$+ \frac{2\chi}{8} \sum_{\alpha,\beta,\gamma=\pm 1} \Big[\eta_z^2 \tilde{D}_{45} \left(\dot{w}_1^{I,J,K+\gamma} - \dot{w}_1^{I,J,K} \right) \Big], \qquad (22)$$

$$w_3^{I,J,K,t+1}(D) = -\frac{2\chi}{8} \dot{w}_3^{I,J,K} \sum_{\alpha,\beta,\gamma=\pm 1} \Big[\eta_x^2 \tilde{D}_{55} + \eta_y^2 \tilde{D}_{44} + \eta_z^2 \tilde{D}_{33} \Big]$$

$$+ \frac{\chi}{8} \sum_{\alpha,\beta,\gamma=\pm 1} \Big[2\eta_x^2 \tilde{D}_{55} \dot{w}_3^{I+\alpha,J,K} + 2\eta_y^2 \tilde{D}_{44} \dot{w}_3^{I,J+\beta,K}$$

$$+ 2\eta_z^2 \tilde{D}_{33} \dot{w}_3^{I,J,K+\gamma} \Big]$$

$$+ \frac{\chi}{8} \sum_{\alpha,\beta,\gamma=\pm 1} \Big[\beta\gamma\eta_y\eta_z \tilde{D}_{44}$$

$$\times \left(\dot{w}_2^{I,J+\beta,K+\gamma} - \dot{w}_2^{I,J,K} - \dot{w}_2^{I,J+\beta,K} + \dot{w}_2^{I,J,K+\gamma} \right) \Big]$$

$$+ \frac{\chi}{8} \sum_{\alpha,\beta,\gamma=\pm 1} \Big[\alpha\gamma\eta_x\eta_z \tilde{D}_{55}$$

$$\times \left(\dot{w}_1^{I+\alpha,J,K+\gamma} - \dot{w}_1^{I,J,K} - \dot{w}_1^{I+\alpha,J,K} + \dot{w}_1^{I,J,K+\gamma} \right) \Big]$$

$$+ \frac{\chi}{8} \sum_{\alpha,\beta,\gamma=\pm 1} \Big[\beta\gamma\eta_y\eta_z \tilde{D}_{45}$$

$$\times \left(\dot{w}_1^{I,J+\beta,K+\gamma} - \dot{w}_1^{I,J,K} + \dot{w}_1^{I,J,K+\gamma} - \dot{w}_1^{I,J+\beta,K} \right) \Big]$$

$$+ \frac{\chi}{8} \sum_{\alpha,\beta,\gamma=\pm 1} \Big[\alpha\gamma\eta_x\eta_z \tilde{D}_{45}$$

$$\times \left(\dot{w}_2^{I+\alpha,J,K+\gamma} - \dot{w}_2^{I,J,K} + \dot{w}_2^{I,J,K+\gamma} - \dot{w}_2^{I+\alpha,J,K} \right) \Big]$$

$$+ \frac{2\chi}{8} \sum_{\alpha,\beta,\gamma=\pm 1} \Big[\alpha\beta\eta_x\eta_y \tilde{D}_{45} \left(\dot{w}_3^{I+\alpha,J+\beta,K} - \dot{w}_3^{I,J,K} \right) \Big],$$

$$(23)$$

where $\tilde{D}_{11} = D_{11}(I + \alpha, J + \beta, K + \gamma)$ represents the viscosity coefficient D_{11} in one of the eight cells surrounding the target point C depending on the choice of (α, β, γ) from $(+1, -1)$; similar expressions hold for the other damping terms. It should be noted that all the first-order time derivatives in Eqs. (21)–(23) are given in the abbreviated form for the sake of presentation compactness. All of these abbreviated expressions should be replaced by the left-sided second-order finite difference transformation. For instance, $\dot{w}_1^{I,J,K}$ should enter the computation in its finite difference form as

$$\dot{w}_1^{I,J,K} = \frac{3w_1^{I,J,K} - 4w_1^{I,J,K,t-1} + w_1^{I,J,K,t-2}}{2\Delta t}. \qquad (24)$$

These extended damping contributions with the stiffness contribution can now capture the guided wave propagation in composite laminates with anisotropic damping effects of an arbitrary stacking orientations.

2.3.1. *UM-LISA Solution Procedure and Parallel CUDA Implementation*

In modeling waveguides with complex structural features, one obstacle is the generation of the computational grid and the allocation of material properties to the cells. The LISA equations require the identification of 18 neighboring nodes and the material properties in the eight surrounding cells to each target node.

Figure 6 shows the UM-LISA solution procedure. To achieve this goal, a commercial FEM preprocessor (ANSYS 14) is integrated seamlessly in the framework to define the structural geometry, carry out discretization, and allocate material properties to each element. This allows the setup of a target problem with complex composite material features such as lamination angles and geometric features such as stiffeners, rivet holes, structural damage, etc. The connectivity and material allocation are converted from FEM to LISA format by providing the numbering

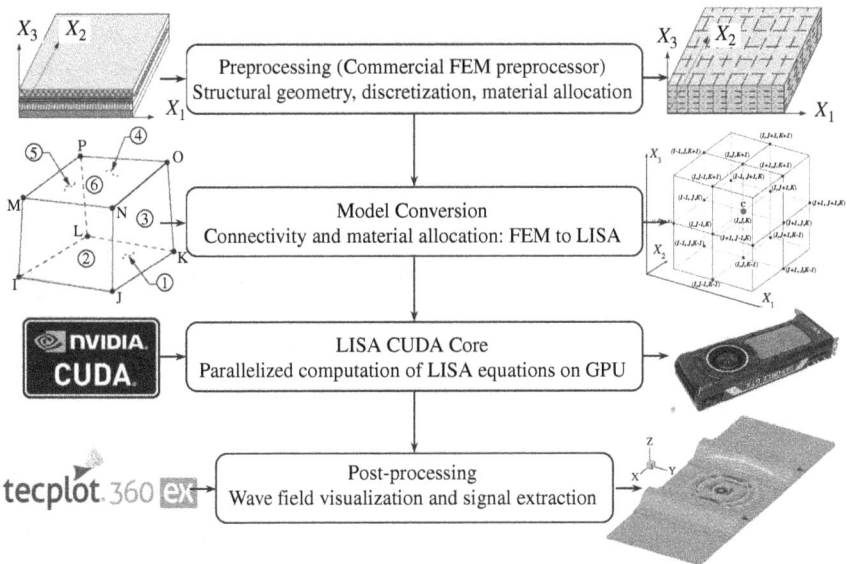

Figure 6. UM-LISA solution procedure.

and properties of the neighboring nodes and cells surrounding each computational node. The model connectivity and material properties are then fed into the LISA core, where the iterative equations of LISA are computed. The time-marching LISA equations are parallelized using CUDA technology and executed on NVDIA GPUs. After the solution is obtained, tecplot 360 ex is utilized to perform wavefield visualization and signal extraction.

Utilizing graphic cards for scientific computation is drawing more and more attention from the engineering community. CUDA has enabled to program and execute general purpose scientific computations on powerful NVIDIA graphic cards. The computational capability of GPU has enjoyed a revolutionary progress during the past few years, surpassing that of CPU.[27,28] It should be noted that a single CPU core still possesses much faster speed than a single CUDA core. The superb computational capability of GPU resides in the following two facts:

(1) GPU holds a massive number of CUDA cores. The smallest computational unit on a CUDA core is called a thread, on which a kernel (function/algorithm) will be executed. Each CUDA core holds multiple threads. Therefore, a GPU device possesses considerably large number of independent computational units. For instance, the GeForce Titan used in this study has 14 multiprocessors, with 192 CUDA cores in each of them. The total number of CUDA cores is 2,688. The number of concurrent threads reaches 28,672, given 2,048 concurrent threads per multiprocessor.
(2) CUDA cores and threads work concurrently in parallel, rather than in series as CPU cores do. This aspect endows great advantage to GPU when operations need to be carried out on each element of an array. A typical C/C++ program will treat array operation element by element using a loop. On the other hand, CUDA assigns the element operation to the threads. Each thread will handle one element. All these threads will run the kernel concurrently. Thus the operations per unit time on a GPU device far surpass that on a CPU.

It should be noted that not all kinds of computations are suitable for GPU. An algorithm can be greatly expedited only when it meets the following two requirements: (1) the algorithm should be massively parallel and (2) the target problem should be computationally intensive. LISA satisfies these two aspects. For the first aspect, LISA is massively parallel, because the computation of one node only depends on the solutions of its 18 neighboring points at the previous three time steps. Thus, the behavior of each node, at current time step, is independent from the others, i.e., the computation of each node can be carried out individually in parallel. For the second aspect, the wave propagation simulation task requires dense discretization of the structure, resulting in a computationally intensive problem. In order to take advantage of the nice parallelizable feature of LISA and the superb computational capability of powerful GPU devices, the LISA procedure was implemented using CUDA. All the parameters are first established in the host memory (RAM). Then a copy of these parameters is sent to the device memory (GPU global memory) for it to be processed. The computation of each node is assigned to a functional thread, i.e., each thread will gather the displacements of its 18 neighboring nodes at previous three time steps, process the material properties in the eight surrounding cells, and execute the kernel to compute the displacement of this node at the current time step. Since one of the bottlenecks of a CUDA program is data transfer between the device memory and host memory, only the required step results are gathered (every 20–30 steps depending on the frequency of the propagating waves) from the GPU to the CPU to minimize such data transfer.

3. Attenuated Guided Wave Propagation in Composite Structures

This section presents numerical simulation results for guided wave propagation on composite laminates, compared with the experiments with SLDV. These results demonstrate the capability of the LISA framework to accurately model guided wave generation, propagation and attenuation in composite structures.

3.1. *Experiments with Scanning Laser Doppler Vibrometry*

SLDV has been widely used to visualize wave propagation in solids for SHM research. Figure 7 shows the experimental setup using SLDV. A Polytec PSV-400-M4 system was used to measure the out-of-plane particle velocity of the scanning surface on the composite laminates. An Agilent 33522A function generator was used to generate three-cycle tone burst signals at various center frequencies. A Krohn-hite 7500 amplifier was used to amplify the excitation to 10 V peak-to-peak (Vpp).

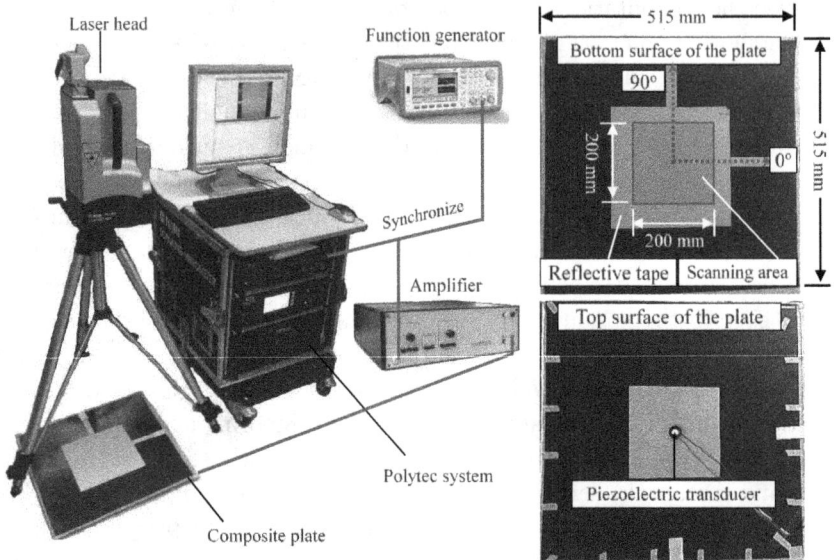

Figure 7.   Experimental setup using scanning laser Doppler vibrometry.

Figure 7 also shows the unidirectional composite plate used in this study. A 12.8-mm diameter and 0.23-mm thick circular piezo wafer was bonded on the center of the plate surface. The bottom surface was used as the scanning surface for SLDV experiments. Reflective tape was used to enhance the signal quality. Under electrical excitation, the piezo wafer will generate mechanical guided waves in the composite plate. The guided waves propagate with an out-spreading pattern, undergo dispersion, experience viscous dissipation and are finally picked up by the SLDV. The multilayered composite specimens used in this study were unidirectional $[0]_{12T}$ and cross-play $[0/90]_{3S}$ CFRP composite plates. They were manufactured from 0.125-mm thick pre-impregnated composite tape made from IM7 fibers and Cycom 977-3 resin. The material mechanical properties associated with our specimen are given in Table 1 and the piezoelectric properties are given in Table 2.

The method of obtaining viscosity coefficients of solids itself is a challenging branch of study. Available data is also limited for CFRP composite materials in current literature. Thus, in this study, the viscosity coefficients were obtained by updating our LISA model toward the experimental measurements. The chosen viscosity matrix coefficient for the unidirectional CFRP lamina is given in Table 3.

Table 1. Mechanical properties of different materials in the experiment and simulations.

Mech. prop.	IM7/977-3	PZT 5A
E_{11} (GPa)	147.00	60.98
E_{22} (GPa)	9.80	60.98
E_{33} (GPa)	9.80	53.19
ν_{12}	0.41	0.35
ν_{13}	0.41	0.44
ν_{23}	0.69	0.44
G_{12} (GPa)	5.3	22.57
G_{13} (GPa)	5.3	21.05
G_{23} (GPa)	3.31	21.05
ρ (kg/m^3)	1558	7750

Table 2. Transducer piezoelectric properties.

Piezo. prop.	PZT 5A
e_{15} (C/m^2)	12.29
e_{25} (C/m^2)	12.29
e_{31} (C/m^2)	-5.35
e_{32} (C/m^2)	-5.35
e_{33} (C/m^2)	15.78
κ_{11} (nF/m)	8.13
κ_{22} (nF/m)	8.13
κ_{33} (nF/m)	7.32

Table 3. Measured IM7/977-3 viscosity coefficients of unidirectional CFRP lamina.

Viscosity coefficients	D_{11}	D_{22}	D_{33}	D_{44}	D_{55}	D_{66}
Unit (Pa·s)	328.6	525.7	525.7	722.8	394.3	394.3

3.2. *Guided Waves in Composite Laminates*

Dispersion relations for the composite laminates are obtained using the semianalytical finite element (SAFE) method in SAFE-DISPERSION software.[29] Figure 8 shows the group velocity curves and directivity wave curves in the 1.5-mm thick $[0]_{12T}$ CFRP composite specimen. Figure 8(a) shows the group velocity curves in 0° direction (along the fibers). It can be observed that, at relatively low frequency range, only three fundamental wave modes exist, namely fundamental symmetric mode (S0) fundamental SH mode (SH0) and fundamental antisymmetric mode (A0). The S0 mode is

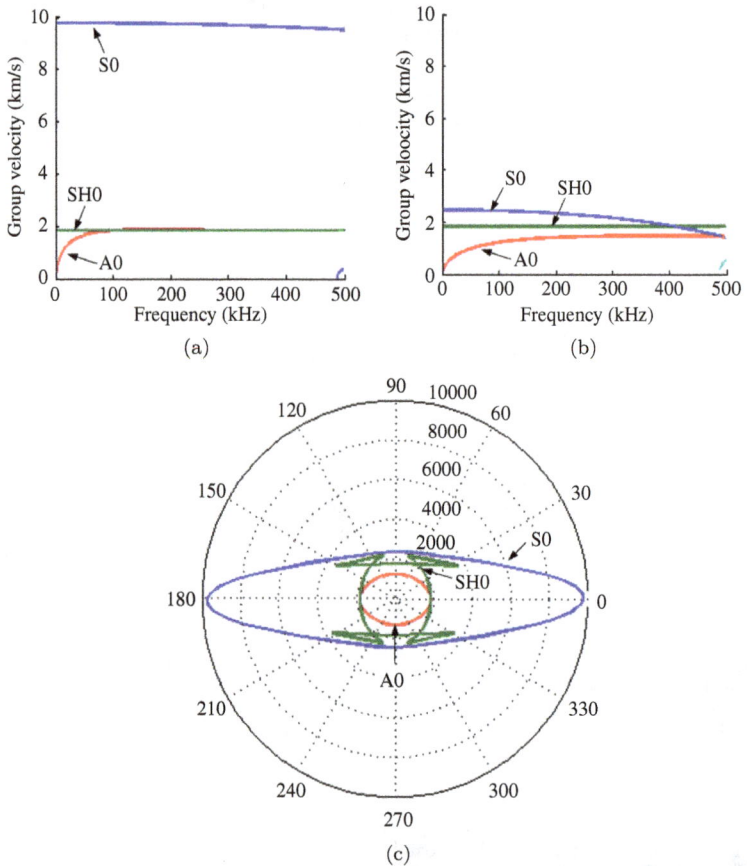

Figure 8.   Guided wave modes in the 1.5-mm thick $[0]_{12T}$ CFRP composite specimen:
(a) group velocity curves along fiber direction; (b) group velocity curves in the fiber
perpendicular direction; (c) wave curves (group velocity directivity curves) at 120 kHz.

not heavily dispersive and has a very high propagation speed. The A0 mode,
on the other hand, is highly dispersive and has the lowest wave speed. The
SH0 mode is non-dispersive at the given frequency range and possesses a
wave speed in between the other two. Figure 8(b) shows the group velocity
curves in the 90° direction. It should be noted that, compared with the
situation in the 0° direction, the S0 mode becomes more dispersive and
has a much lower propagation speed. The SH0 wave speed does not change
much, while the A0 wave speed becomes much lower. Figure 8(c) shows the
directivity plot of the wave curves (group velocity) at 120 kHz. The wave
curves represent the spatial wave propagation pattern. It can be observed

that both S0 and A0 have the highest propagation speed along the fiber direction, while their wave speeds perpendicular to the fiber direction are much lower, exhibiting elliptical wavefronts. The SH0 wave curve shows the well-known self-crossing behavior. The SH0 wave speed in the fiber direction approaches that of the A0 mode, which may make them hard to be distinguished. However, along other propagation directions, they are easily identifiable. This phenomena will be shown in a later section.

Figure 9 shows the group velocity curves and directivity wave curves in the 1.5-mm thick cross-ply $[0/90]_{3S}$ CFRP composite specimen. Figures 9(a)

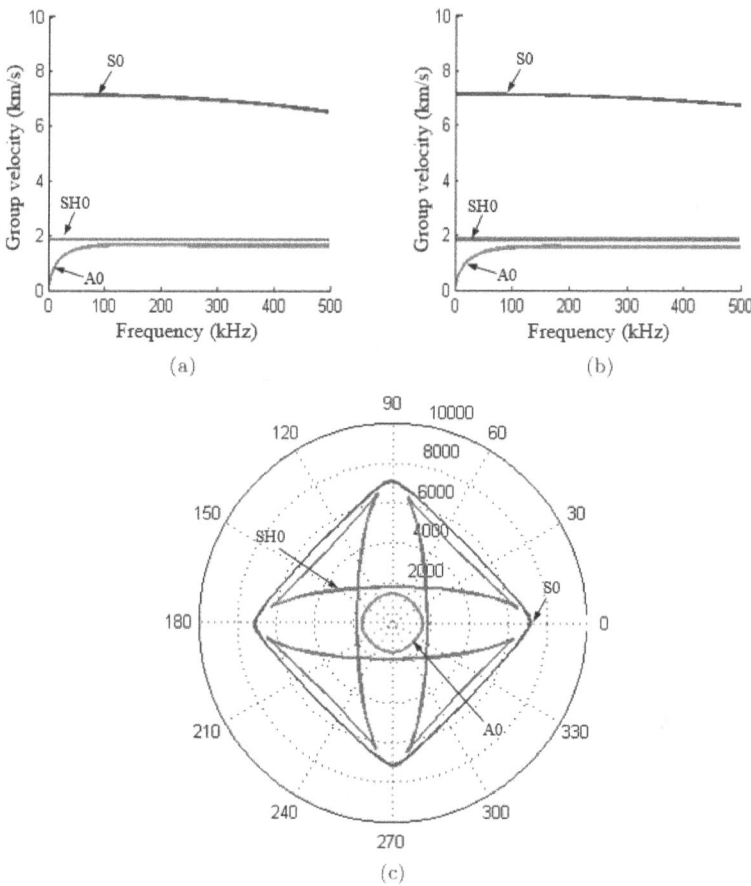

Figure 9. Guided wave modes in the 1.5-mm thick $[0/90]_{3S}$ composite plate: (a) group velocity curves along $0°$ direction; (b) group velocity curves along $90°$ direction; (c) wave curves (group velocity directivity curves) at 75 kHz.

and 9(b) present the group velocity curves in the 0° and 90° directions, respectively. It can be observed that, in both directions, three fundamental wave modes exist: S0, A0 and SH0. The S0 mode has the highest wave speed and A0 the lowest speed, with SH0 in between the two. At relative low frequency range, the wave speeds in these two directions are approximately the same. Small differences begin to show as the frequency increases. This means higher frequency waves become more and more sensitive to the detailed stacking layer structural changes. Figure 9(c) shows the directivity wave curves of the cross-ply composite plate at 75 kHz. It can be noticed that the wave speeds are highest along the fiber directions (0° or 90°).

3.3. *Validation of Guided Wave Generation of Hybrid LISA Solution Against SLDV*

This section presents the validation of the guided wave excitability using the hybrid model. Guided wave generation from a circular piezo wafer on unidirectional and cross-ply composite plates are modeled. The wave generation patterns are compared with experimental measurements from SLDV.

Figure 10 shows the comparison between the experimental and the hybrid LISA simulation results in the unidirectional composite plate at 75 kHz. It can be noticed that the wave pattern obtained using the hybrid LISA agrees very well with the experimental result from SLDV. Strong A0 mode guided waves are generated along the fiber direction, while the wave energy perpendicular to the fiber direction is relatively weak. The wave signals 15 mm away from the transducer center were selected and enveloped. The maximum amplitude of the enveloped signals is plotted in Fig. 10(c) to compare the excitation results. The root mean square (RMS) error between the LISA prediction and the experimental data is shown in Fig. 10(d). In general, a good match between the experiment and hybrid LISA can be observed. It should be noted that LISA assumes the ideal bonding of PZT transducer with a uniform thickness plate. The error may be attributed to the variation in bonding layer quality and manufactured plate thickness.

Figure 11 shows the comparison between the experimental and the hybrid LISA simulation result in the unidirectional composite plate at 120 kHz. The hybrid LISA result agrees well with the experiment. It can be observed that strong A0 mode waves are generated perpendicular to

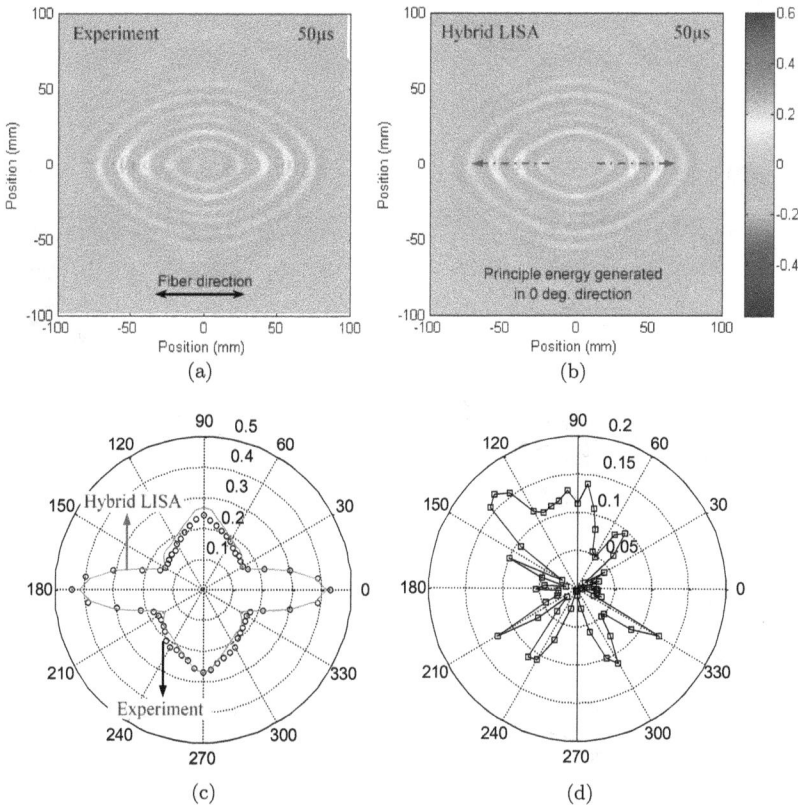

Figure 10. Comparison of wave generation pattern in the unidirectional composite plate at 75 kHz: (a) experimental result; (b) hybrid LISA result; (c) wave amplitude directivity plot; (d) RMS error of wave amplitude directivity plot.

Note: The principal wave energy is generated along the fiber direction.

the fiber direction, which is distinctively different from the wave generation behavior shown at 75 kHz (Fig. 10). Along the fiber direction, the A0 mode reaches the rejection frequency, where the wave amplitude is diminished.[30] In addition, the S0 and the SH0 modes can be easily identified in the wavefront pattern by comparing with the wave curves in Fig. 8(c). They propagate in the laminate with different wave speeds and wavelengths. The excitation amplitude directivity and the RMS error are shown in Fig. 11(c) and Fig. 11(d). The 0° direction has relatively big error, while the agreement in other directions is achieved well.

Figure 12 presents the wave pattern and excitation amplitude comparison in the cross-ply composite plate at 75 kHz. The hybrid LISA

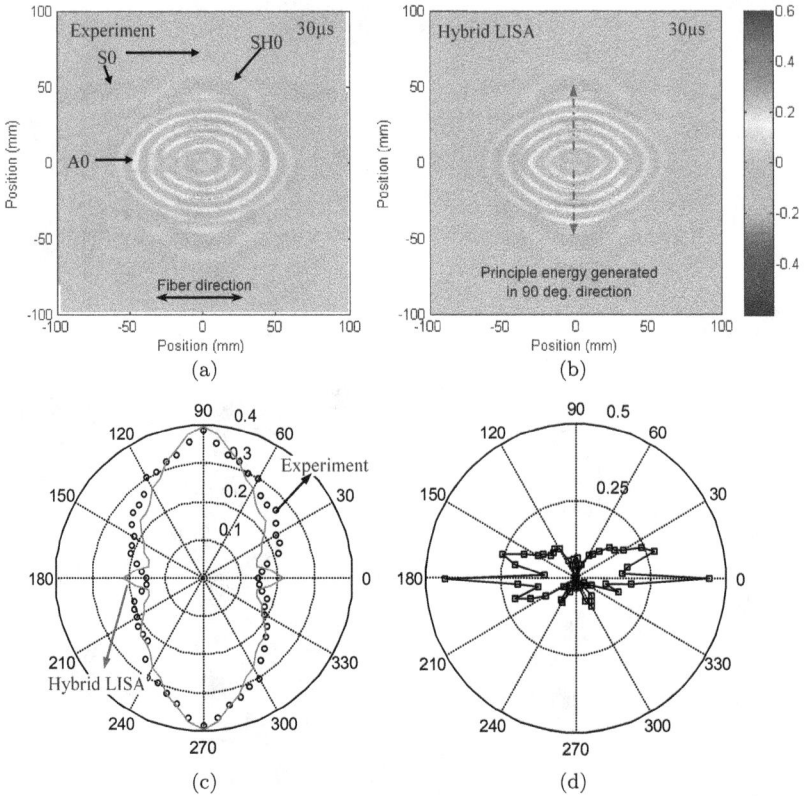

Figure 11. Comparison of wave generation pattern in the unidirectional composite plate at 120 kHz: (a) experimental result; (b) hybrid LISA result; (c) wave amplitude directivity plot; (d) RMS error of wave amplitude directivity plot.

Note: The principal wave energy is generated perpendicular to the fiber direction.

simulation result shows good agreement with experimental measurement. The A0 wave speed has nearly the same propagation speed along the 0° and 90° directions. The excitation amplitude is also very close in these two directions as shown in the directivity plot in Fig. 12(c). The principal energy is generated along the 0° and 90° directions. Hybrid LISA solution compares well with SLDV experimental result. Fig. 12(c) shows the wave amplitude directivity plot. It can be observed that the experimental data is not perfectly symmetric due to the variation of PZT bonding and composite manufacturing. This is also reflected in the RMS plot in Fig. 12(d).

Figure 13 presents the wave pattern and excitation amplitude comparison in the cross-ply composite plate at 120 kHz. At this frequency,

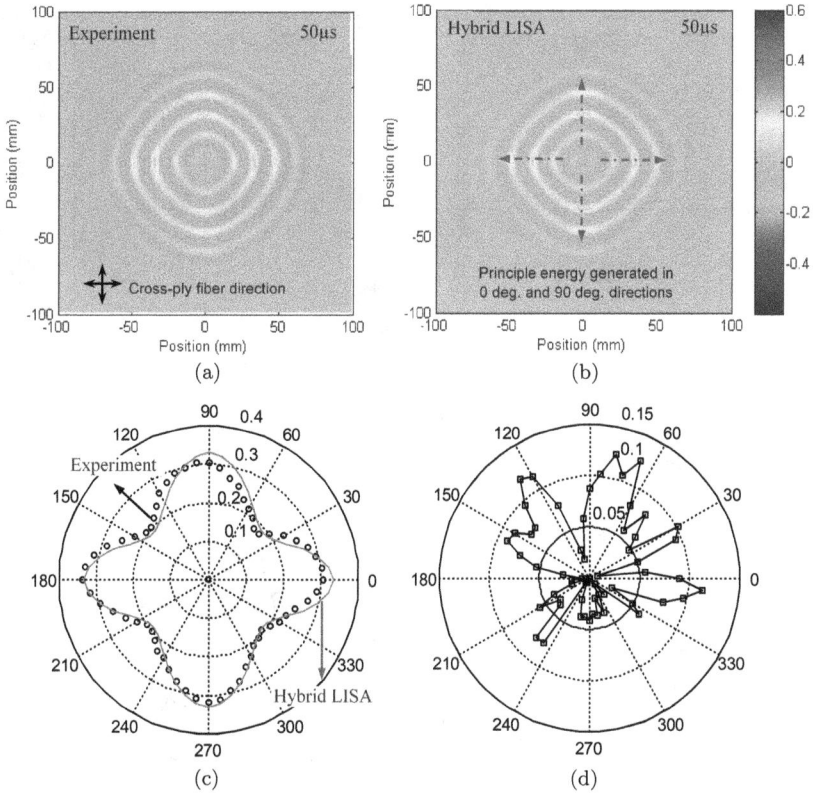

Figure 12. Comparison of wave generation pattern in the cross-ply composite plate at 75 kHz: (a) experimental result; (b) hybrid LISA result; (c) wave amplitude directivity plot; (d) RMS error of wave amplitude directivity plot.

Note: The principal wave energy is generated equally along the fiber directions.

the S0 and SH0 modes can be identified. It can be noticed that the wave energy is generated more evenly than the 75 kHz case in all the propagation directions for A0 mode. As the frequency moves higher, the variation in the experiment becomes more obvious. Figures 13(c) and 13(d) show a relatively good match between the LISA prediction and experimental measurement.

3.4. Comparison of Anisotropic Damping Effects in Guided Wave Propagation

Figure 14 shows the comparison of waveforms at various distances from the transducer along 0°, 45° and 90° directions. All the signals are

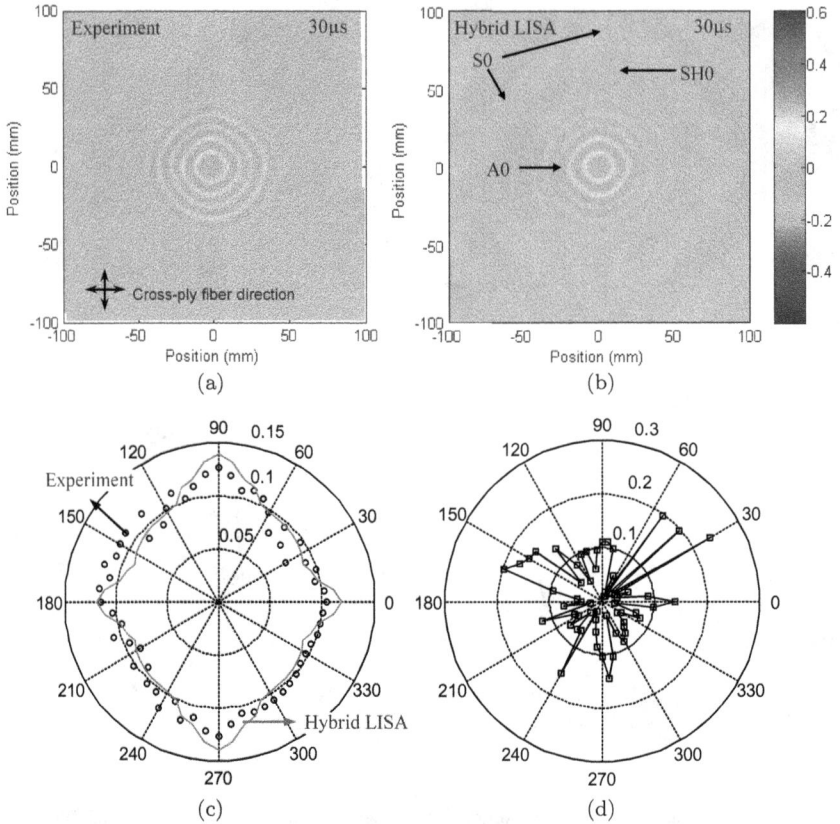

Figure 13.   Comparison of wave generation pattern in the cross-ply composite plate at 120 kHz: (a) experimental result; (b) hybrid LISA result; (c) wave amplitude directivity plot; (d) RMS error of wave amplitude directivity plot.

Note: The principal wave energy is generated equally along the fiber directions.

normalized to the maximum oscillation at the transducer. The 75-kHz signals in the unidirectional composite plate at 20, 40, 60, 80 and 100 mm away from the transducer are plotted. The waveforms obtained by the hybrid LISA simulation achieved remarkable agreement with experimental measurements. It can be observed that the signals near the transducer have higher amplitudes. In all three examined propagation directions, the wave amplitudes attenuate as the sensing locations move further away from the wave source. In the 0° direction, along the fiber, strong A0 wave mode can be observed. However, the amplitude of the S0 wave is

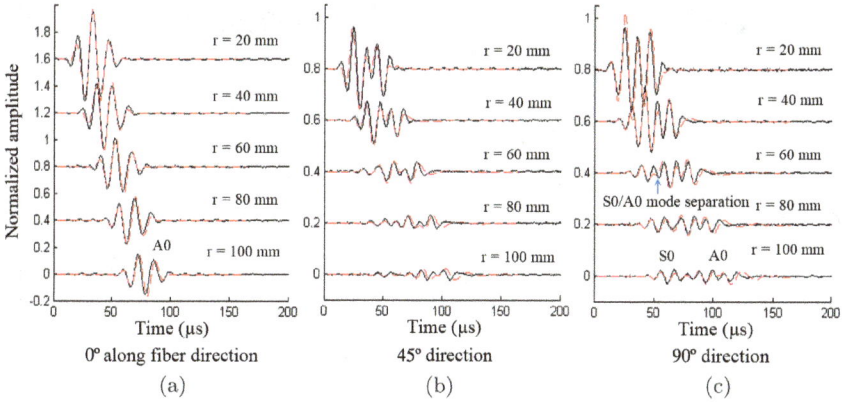

Figure 14. Waveform comparison at various distances from the transducer along 0°, 45° and 90° directions. (Solid line: experimental measurement; dashed line: hybrid LISA simulation.)

very weak. It should be noted that the S0 mode may undergo large in-plane motion, while the SLDV only measures the out-of-plane motion, which is relative small for the S0 mode in the 0° direction. On the other hand, in 90° direction, strong S0 amplitude was picked up and can be distinguished. At $r = 60$ mm, mode separation takes place, where the fast S0 waves overtake the slow A0 waves, and leads the propagating waves. At $r = 100$ mm, the S0 and A0 wave packets are clearly separated. By comparing the wave amplitude change in various directions, one may notice that the wave attenuation is much stronger in the 45° and 90° directions than the 0° direction. This shows that the extended viscoelastic LISA formulation can successfully capture the anisotropic damping effects in the orthotropic composite laminate.

The wave amplitude attenuation in this study can be attributed to three main aspects: (1) energy distribution due to the out-spreading propagation pattern; (2) spatial elongation of wave packets due to dispersion; and (3) energy dissipation due to damping. Conventional LISA formulation can capture the first two effects, while it did not take into account the third aspect. Figure 15 presents the comparison of the new LISA with damping and the conventional LISA without damping against the experimental measurement. It shows that both conventional LISA and the extended LISA can depict the out-spreading wave pattern and dispersive spatial elongation. But the difference appears along 90° direction, where the wave amplitude stayed strong in the conventional LISA result without damping.

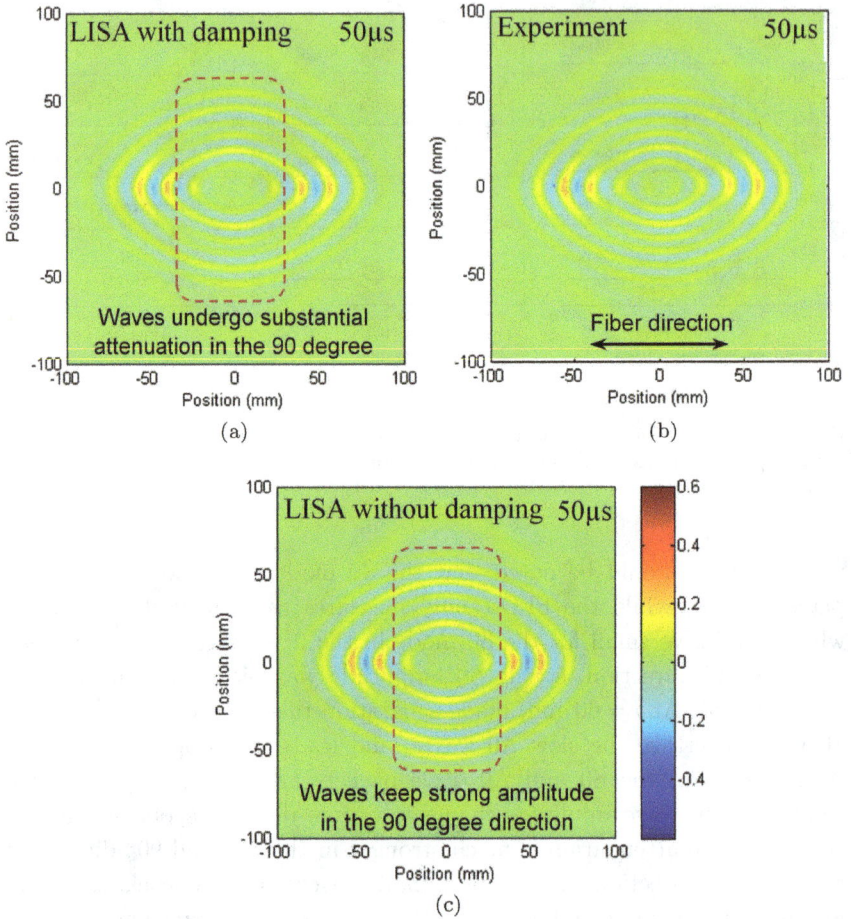

Figure 15. Wave pattern in a $[0]_{12T}$ composite plate: (a) LISA simulation with extended damping effect; (b) experimental measurement; (c) LISA simulation without damping effect.

In comparison, the extended LISA formulation with damping effect agrees much better with the experiment, where obvious wave attenuation was noticed along 90° direction.

Figure 16 shows the wave amplitude change along three propagation directions. The wave amplitudes are normalized to the maximum oscillation at the transducer. The results from LISA with damping and without damping are compared with experimental data. Again, a good match was achieved with the new formulation with anisotropic damping effects. The

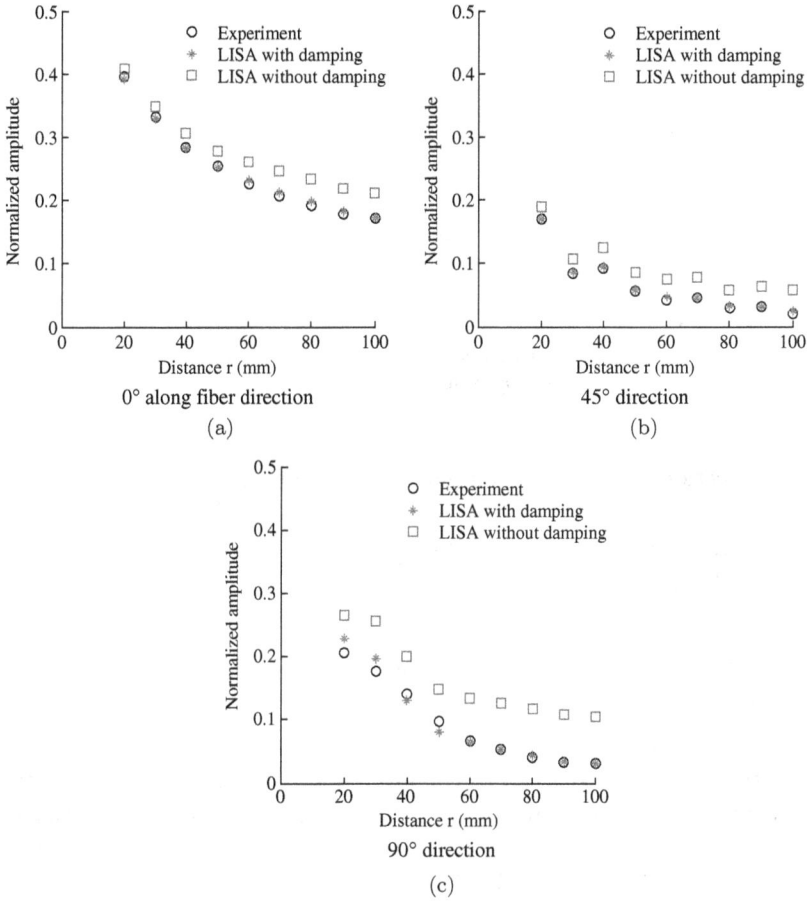

Figure 16. Improvement of amplitude agreement after adding damping effects. Wave attenuation comparison along various propagation directions the in $[0]_{12T}$ composite plate case.

difference between the LISA results with and without damping shows the amount of attenuation attributed to damping. It is apparent that damping has the highest influence on the 90° propagation direction.

To further demonstrate the improvement obtained with the extended LISA formulation, the signal result at (0 mm, 100 mm) location is presented in Fig. 17. The signal amplitudes are normalized to the maximum oscillation at the transducer. It can be seen that the conventional LISA overestimates the A0 mode amplitude without considering the directional structural damping. On the other hand, the extended viscoelastic LISA formulation

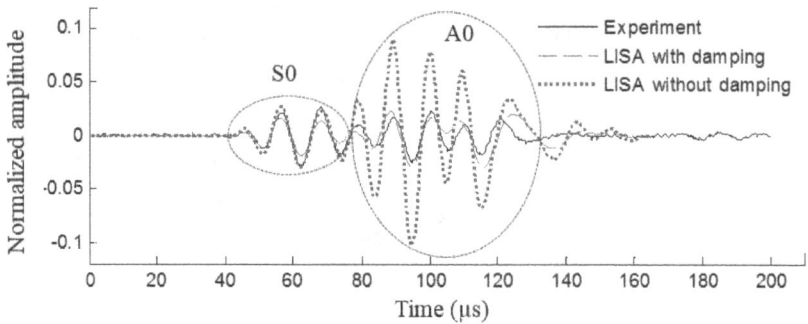

Figure 17. Simulation signal with and without damping effect and the measured signal at a point 100 mm away from the source normal to the fibers in a $[0]_{12T}$ composite laminate.

takes directional structural damping into account. It should also be noted that the damping effects have more influence on the A0 mode than the S0 mode. This is related to the motion characteristics of their mode shapes and wavelengths.

4. Artificial Absorbing Boundary for Wave Propagation Simulation

To ensure computational accuracy, strict rules of spatial and temporal discretization need to be obeyed. Thus, the propagation of high-frequency short-wavelength waves over long distances may become computationally prohibitive because of the very fine mesh and very small time step required to ensure the validity of the simulated wave signals, especially in interaction with structural flaws and damage.[31]

In order to make the computational burden manageable, one notices that the numerical model is mainly necessary in the study of the scattering interaction between waves and structural damage because the LISA approach allows detailed modeling of the damage geometry. And the use of large plate models has only been justified by the need to avoid boundary reflections contaminating the scatter signal. Hence, researchers have developed the concept of absorbing boundary condition (ABC), which would eliminate the unwanted boundary reflections and allow the use of a finite size numerical model to simulate infinite medium conditions. The benefit of ABC approach is that computational resources can be focused on the region of interest without modeling the redundant outside domain merely for the purpose of avoiding boundary reflections.

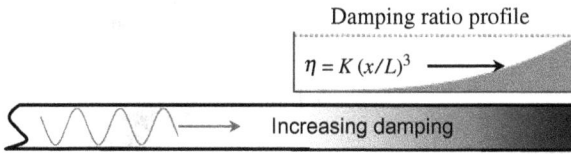

Figure 18.   Absorbing layers by increasing damping (ALID) method.

Figure 18 presents the absorbing layers by increasing damping (ALID) method, which has been widely used in FEM simulations. It extends the boundary using several layers of absorbing elements with gradually varying properties. The damping mechanism contributes to the imaginary part of the wavenumbers, which results in the attenuation of the wave amplitudes along the propagation path. The ALID method adopts increasing damping along wave propagation to absorb incident waves, accompanied by impedance mismatches between successive layers[32–36] The impedance-mismatch reflections may be minimized by optimizing the damping properties. A more recent contribution to ALID-type method was given by Pettit *et al.*[37] who developed a stiffness reduction method (SRM) to further optimize its performance. Drozdz *et al.*[38] provided the guideline of choosing the damping profile, and later investigations also demonstrated its effectiveness.[39]

ALID method is adopted in the LISA numerical framework due to the readiness of its implementation. Further details of ALID method implementation in current UM-LISA framework can be found in Ref. [40]. Figure 19 shows the LISA simulation results of wave propagation in a strip specimen. An incident wave is sent to interactive with (1) ordinary reflective boundary and (2) artificial ALID absorbing boundary. It can be observed that obvious reflections happened at the reflective strip end, while the ALID strip boundary effectively absorbed the incoming waves, i.e., the simulation acts as if the wave propagated into an infinite media.

Taking advantage of the ALID method, small-sized local LISA models can be constructed to efficiently study the interaction between guided waves and structural damage, as shown in Fig. 20(a). LISA can simulate arbitrary damage features by defining the grid and material properties. The red circle represents the excitation nodes, which can generate incident waves with arbitrary impinging angle toward the damage. The blue circle shows the sensing boundary to pick up the signal at an arbitrary scattering angle from the damage. The outer region is extended with ALID boundary to eliminate reflections. Such a local LISA model can save considerable

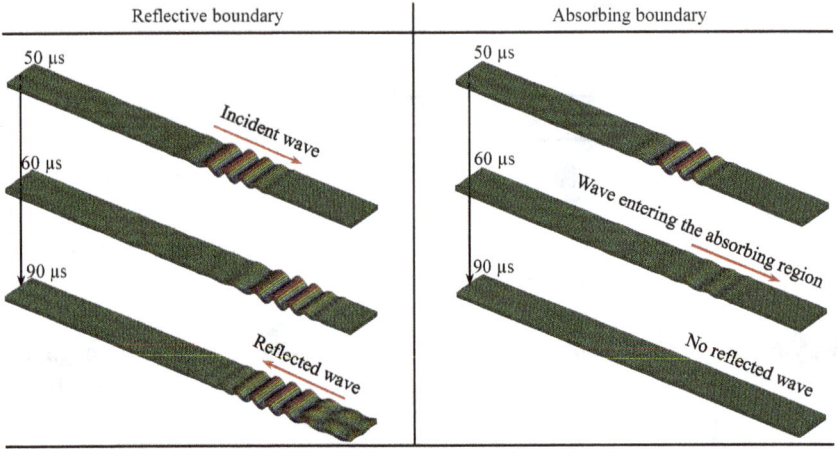

Figure 19.   LISA ALID absorbing boundary eliminates reflected waves.

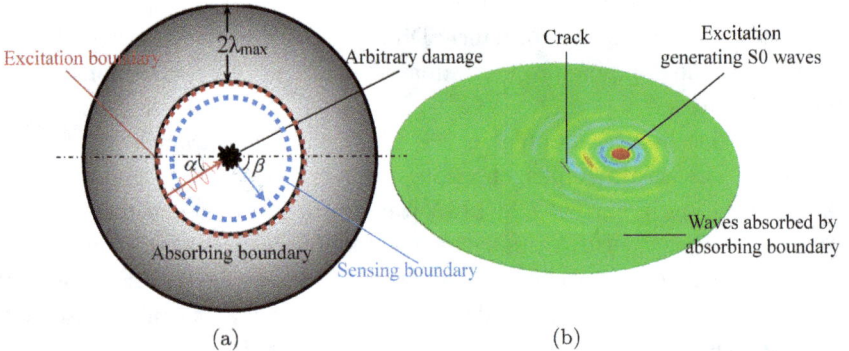

Figure 20.   (a) Local LISA model with absorbing boundary to study wave damage interaction; (b) Local LISA simulation of wave interaction with a crack damage.

computational time and resource by focusing the computational efforts on the target problem of wave damage interaction without the influence from boundary reflections. Figure 20(b) presents the simulation result of S0 wave interaction with a crack using the local LISA model. It should be noted that selective wave mode generation can be achieved through coupling/decoupled excitation profiles with guided wave mode shapes, which allows the study of the sensitivity of various guided wave modes to a certain type of damage. At the sensing locations, mode decomposition can be done to study the mode conversion effects from wave damage interaction. The waves generated by the excitation points propagate into the area of interest, interact with the crack damage, be picked up at the sensing points, and are finally absorbed by the ALID region, eliminating the influence from the boundary reflections.

This small-sized model technique with absorbing boundary condition has been widely investigated in FEM for linear wave damage interaction in metallic structures.[41,42] And the frequency domain solution proves to be very efficient[43,44] However, the study of nonlinear wave damage interaction (such as wave interaction with fatigue crack, composite delamination, composite impact damage, etc.) is only possible using transient dynamic solutions. The highly efficient parallel feature makes LISA an excellent choice to attach such problems. And the small-size local LISA model with ALID absorbing boundaries possesses profound potential in such applications.

5. Dealing with Complexity: Guided Wave Interaction with Damage and Structural Features

The interaction of guided waves with structural features and damage is of great interest in SHM/NDE research. In this section, a 1.5-mm thick $[0/90]_{3S}$ cross-ply composite plate with stiffeners and delamination is modeled. Figure 21 shows the schematic of the stiffened composite plate and

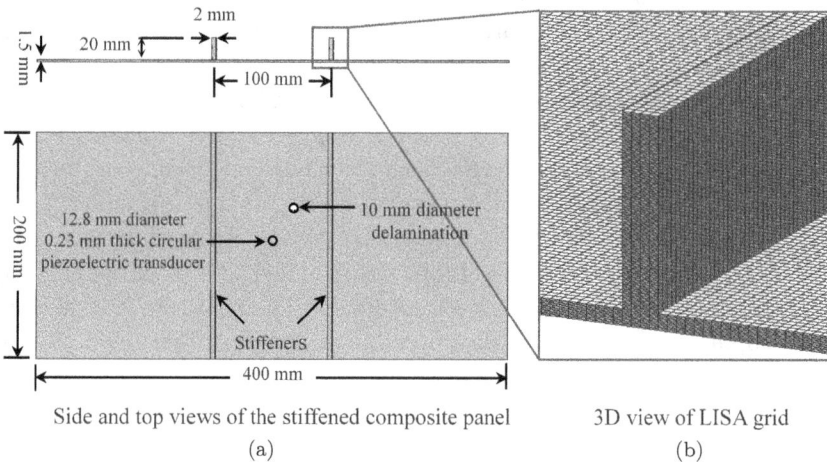

Side and top views of the stiffened composite panel   3D view of LISA grid
(a)   (b)

Figure 21.   Schematic of the blade-stiffened composite panel with delamination and the LISA grid.

the LISA grid. The plate dimension is 400 mm × 200 mm. Two 2-mm thick, 20-mm wide stiffeners are integrated on the plate, 100 mm away from each other. A 10-mm diameter circular shape delamination is placed at (20 mm, 30 mm) with respect to the plate center. Delamination is modeled by setting the corresponding middle layer material properties to those of air. It should be noted that the clapping nonlinear interaction is not considered in this case. A 12.8-mm diameter, 0.23-mm thick circular piezoelectric actuator is bonded on the top surface at its center and sends out a 200-kHz 3.5-cycle tone burst wave. The guided waves generated by the transducer propagate along the structure and interact with the stiffeners and delamination.

There are 12 layers of cells through the thickness, each one of them representing a lamina. It should be noted that a thin layer of air cells was implemented surrounding the structure to simulate free boundary conditions. The in-plane direction has a mesh size of 1 mm. The final LISA grid has 1,551,935 computational nodes and 4,655,805 degrees of freedom.

Figure 22 presents the simulation results illustrating various wave–structure interaction phenomena. The S0 mode has a higher speed along 0 and 90-degree directions. At 21 μs, the interaction between S0 and the stiffener can be clearly observed, with S0 mode converted to A0 mode. The S0 wave has a long wavelength and small out-of-plane motion, while the mode converted A0 mode has a short wavelength and large out-of-plane motion. At 39 μs, the A0 waves interact with the delamination, and diffractions/scattering from the damage can be noticed. At 49 μs, the A0 waves have already propagated through the damage, but trapped modes still exist within the delamination. Also, the transmission and reflection of the A0 waves at the stiffeners can be clearly observed.

It should be pointed out that the LISA CUDA code achieved a remarkable computational speed. With the GeForce Titan NVIDIA GPU, it takes 0.12 s per time step for such a case study. For 6,000 computation steps, the total computational time is only around 10 mins. Such high computational efficiency surpasses far beyond that of conventional FEM simulations for such size problems. It should be pointed out, for conventional FEM simulations, if all the details of laminates are considered, the computation usually becomes prohibitive. On the other hand, GPU-parallelized LISA simulation allows the highly efficient while capturing abundant structural details. The fast prediction of guided wave propagation and interaction with structural features can considerably enhance the SHM system design procedure.

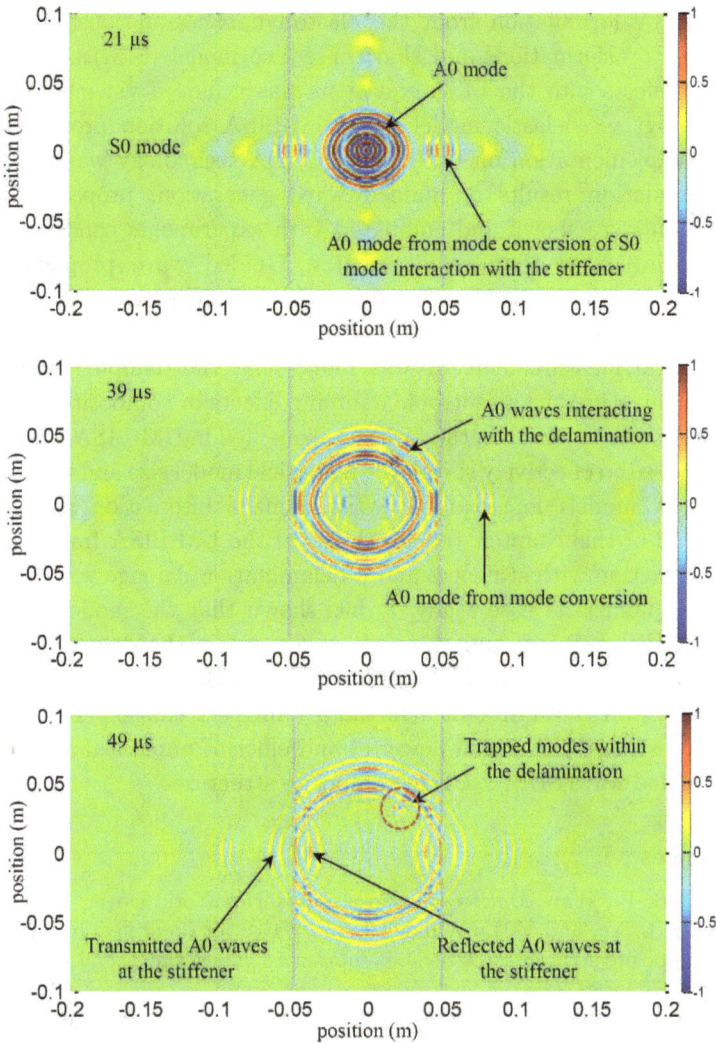

Figure 22. Guided wave propagation in a blade-stiffened cross-ply CFRP composite panel with delamination (0° orientation along the plot horizontal direction).

6. Summary and Concluding Remarks

This chapter presented the LISA modeling technique for the efficient simulation of guided wave generation, propagation and their interaction with damage in composite structures. The chapter started with the basis of UM-LISA formulation from the elastodynamic equations with finite difference transformations and sharp interface model (SIM). Addition of damping effects into the LISA equations was achieved via introducing a Kelvin–Voigt viscoelastic model. The UM-LISA solution procedure and parallel implementation on GPU using CUDA technology was presented. LISA simulation results on guided wave generation, propagation and attenuation were shown and compared with experiments conducted with the scanning laser Doppler vibrometry. Guided wave propagation in unidirectional and cross-ply composite plates were modeled and tested. The LISA simulation wave patterns showed good agreement with the experimental measurements. It was found that the damping model can accurately capture the anisotropic damping effects in composite materials. An ABC based on ALID method was implemented in UM-LISA framework. It can be used to effectively shrink the numerical model size and to construct local LISA models for studying wave damage interaction phenomena. At the end of this chapter, the capability of the UM-LISA framework for wave interaction with stiffeners and a delamination damage was presented through a numerical case study. It has shown that the mode conversion effects at the stiffeners and trapped modes within the delamination can be clearly captured. And it also showed that LISA is capable of dealing with complex structural geometries and damage scenarios. This chapter has demonstrated LISA as a powerful and efficient numerical approach to study guided wave SHM/NDE in composite structures.

References

1. Graff, K. F., *Wave Motion in Elastic Solids*. New York: Dover, 1991.
2. Su, Z., L. Ye, and Y. Lu, Guided Lamb waves for identification of damage in composite structures: A review. *Journal of Sound and Vibration* **295**(3–5) (2006) 753–780.
3. Gresil, M. and V. Giurgiutiu, Prediction of attenuated guided waves propagation in carbon fiber composites using Rayleigh damping model. *Journal of Intelligent Material Systems and Structures* (2014), doi: 10.1177/ 1045389X14549870.
4. Giurgiutiu, V., *Structural Health Monitoring with Piezoelectric Wafer Active Sensors*, 2nd edition. Elsevier Academic Press, Cambridge, Massachusetts, 2014.
5. Moser, F., L. J. Jacobs, and J. Qu, Modeling elastic wave propagation in waveguides with the finite element method. *NDT&E International* **32** (1999) 255–234.

6. Lowe, M., Matrix techniques for modeling ultrasonic waves in multilayered media. *Ultrasonics Ferroelectrics and Frequency Control IEEE Transactions* **42**(4) (1995) 525–542.

7. Obenchain, M. and C. E. S. Cesnik, Producing accurate wave propagation time histories using the global matrix method. *Smart Materials and Structures* **22**(12) (2013) 11.

8. Bartoli, I., A. Marzani, and F. Lanza di Scalea, Modeling wave propagation in damped waveguides of arbitrary cross-section. *Journal of Sound and Vibration* **295**(3–5) (2006) 685–707.

9. Ostachovicz, W., P. Kudela, M. Krawczuk, and A. Zak, Spectral finite element method. In *Guided Waves in Structures for SHM: The Time-Domain Spectral Element Method*. Chichester, UK: John Wiley & Sons, Ltd, 2012.

10. Li, F., H. Peng, X. Sun, J. Wang, and G. Meng, Wave propagation analysis in composite laminates containing a delamination using a three-dimensional spectral element method. *Mathematical Problems in Engineering*, Article ID 659849 (2012), 1–19 doi: 10.1155/2012/659849.

11. Glushkov, E., N. Glushkova, R. Lammering, A. Eremin, and M. Neumann, Lamb wave excitation and propagation in elastic plates with surface obstacles: Proper choice of central frequencies. *Smart Materials and Structures* **20**(1) (2011) 1–11.

12. Delsanto, P., T. Whitcombe, H. Chaskelis, and R. Mignogna, Connection machine simulation of ultrasonic wave propagation in materials I: The one-dimensional case. *Wave Motion* **16**(1) (1992) 65–80.

13. Delsanto, P., R. Schechter, H. Chaskelis, R. Mignogna, and R. Kline, Connection machine simulation of ultrasonic wave propagation in materials II: The two-dimensional case. *Wave Motion* **20**(4) (1994) 295–314.

14. Delsanto, P., R. Schechter, and R. Mignogna, Connection machine simulation of ultrasonic wave propagation in materials III: The three-dimensional case. *Wave Motion* **26**(4) (1997) 329–339.

15. Lee, B. and W. Staszewski, Modelling of Lamb waves for damage detection in metallic structures: Part I. Wave propagation. *Smart Materials and Structures* **12**(5) (2003) 804–814.

16. Lee, B. and W. Staszewski, Modelling of Lamb waves for damage detection in metallic structures: Part II. Wave interactions with damage. *Smart Materials and Structures* **12**(5) (2003) 815–824.

17. Sundararaman, S. and D. Adams, Modeling guided waves for damage identification in isotropic and orthotropic plates using a local interaction simulation approach. *Journal of Vibration and Acoustics* **130**(4) (2008) 1–16.

18. Packo, P., T. Bielak, A. Spencer, W. Staszewski, T. Uhl, and K. Worden, Lamb wave propagation modelling and simulation using parallel processing architecture and graphical cards. *Smart Materials and Structures* **21**(7) (2012) 1–13.

19. Kijanka, P., R. Radecki, P. Packo, W. Staszewski, and T. Uhl, GPU-based local interaction simulation approach for simplified temperature effect modelling in Lamb wave propagation used for damage detection. *Smart Materials and Structures* **22**(3) (2013) 1–16.
20. Nadella, K. and C. E. S. Cesnik, Local interaction simulation approach for modeling wave propagation in composite structures. *CEAS Aeronautical Journal* **4**(1) (2013) 35–48.
21. Sinor, M. Numerical modeling and visualization of elastic waves propagation in arbitrary complex media. In *8th Workshop on Multimedia in Physics Teaching and Learning of the European Physical Society*, Brehova, 2004.
22. Obenchain, M., K. Nadella, and C. E. S. Cesnik, Hybrid global matrix/local interaction simulation approach for wave propagation in composites. *AIAA Journal* **53**(2) (2014) 379–393.
23. Nadella, K. and C. E. S. Cesnik, Piezoelectric coupled LISA for guided wave generation and propagation. In *SPIE Smart Structures and NDE*, San Diego (2013).
24. Nadella, K. and C. E. S. Cesnik, Effect of piezoelectric actuator modeling for wave generation in LISA. In *SPIE Smart structures and NDE*, San Diego, 2014.
25. Shen, Y. and C. E. S. Cesnik, Hybrid local FEM/global LISA modeling of guided wave propagation and interaction with damage in composite structures. In *SPIE Smart Structures and NDE* San Diego, 2015.
26. Saravanos, D. A. and C. C. Chamis, Unified micromechanics of damping for unidirectional fiber reinforced composites. NASA Levis Research Center, Cleveland, Ohio, 1989.
27. Luebke, D. Supercomputing 2009 CUDA tutorial (2009) [Online]. Available: http://gpgpu.org/sc2009.
28. Packo, P., T. Bielak, A. B. Spencer, T. Uhl, W. J. Staszewski, K. Worden, T. Barszcz, P. Russek, and K. Wiatr, Numerical simulations of elastic wave propagation using graphical processing units — Comparative study of high-performance computing capabilities. *Computer Methods in Applied Mechanics and Engineering* **290**(1) (2015) 98–126.
29. Shen, Y. and V. Giurgiutiu, Excitability of guided waves in composites with PWAS transducers. In *41st Annual Review of Progress in Quantitative Nondestructive Evaluation*, AIP Conf. Proc. 1650, 2015, 658–667.
30. Raghavan, A. and C. E. S. Cesnik, Modeling of guided-wave excitation by finite-dimensional piezoelectric transducers in composite plates. In *Proceedings of the 48th AIAA/ASME/ASCE/AHS/ASC Structures, Structural Dynamics, and Materials Conference*, 2007.
31. Gresil, M., Y. Shen, and V. Giurgiutiu, Benchmark problems for predictive FEM simulation of 1-D and 2-D guided waves for structural health monitoring. *Review of Progress in Quantitative Nondestructive Evaluation* **31** (2012) 1835–1842.
32. Israeli, M. and S. Orszag, Approximation of radiation boundary conditions. *Journal of Computational Physics* **41** (1981) 115–135.

33. Castaings, M., C. Bacon, B. Hosten, and M. Predoi, Finite element predictions for the dynamic response of thermo-viscoelastic material structures. *Journal of Acoustical Society of America* **115** (2004) 1125–1133.

34. Castaings, M., C. Bacon, B. Hosten, and M. Predoi, Finite element modeling of torsional wave modes along pipes with absorbing materials. *Journal of Acoustical Society of America* **119** (2006) 3741–3751.

35. Ke, W., M. Castaings, and C. Bacon, 3D finite element simulations of an air-coupled ultrasonic NDT system. *NDT&E International* **42** (2009) 524–533.

36. Semblat, J., L. Lenti, and A. Gandomzadeh, A simple multi-directional absorbing layer method to simulate elastic wave propagation in unbounded domains. *International Journal of Numerical Methods in Engineering* **85** (2011) 1543–1563.

37. Pettit, J., A. Walker, P. Cawley, and M. Lowe, A stiffness reduction method for efficient absorption of waves at boundaries for use in commercial finite element codes. *Ultrasonics* **54** (2014) 1868–1879.

38. Drozdz, M., L. Morreau, M. Castaings, M. Lowe, and P. Cawley, Efficient numerical modelling of absorbing regions for boundaries of guided waves problems. *AIP Conference Proceedings* **820** (2006) 126–133.

39. Shen, Y. and V. Giurgiutiu, Effective non-reflective boundary for Lamb waves: Theory, finite element implementation, and applications. *Wave Motion* (2015), doi: 10.1016/j.wavemoti.2015.05.009.

40. Zhang, H. and C. E. S. Cesnik, A hybrid non-reflective boundary technique for efficient simulation of guided waves using local interaction simulation approach. In *SPIE Smart Structures and NDE*, Las Vegas, 2016.

41. Velichko, A. and P. Wilcox, Post-processing of guided wave array data for high resolution pipe inspection. *Journal of the Acoustical Society of America* **126**(6) (2009) 2973–2982.

42. Velichko, A. and P. Wilcox, A generalized approach for efficient finite element modeling of elastodynamic scattering in two and three dimensions. *Journal of the Acoustical Society of America* **128**(3) (2010) 1004–1014.

43. Velichko, A. and P. Wilcox, Efficient finite element modeling of elastodynamic scattering from near surface and surface breaking defects. In *AIP Conf. Proc.* **1335** (2011) 59, doi: 10.1063/1.3591840.

44. Velichko, A. and P. Wilcox, Efficient finite element modeling of elastodynamic scattering with non-reflective boundary conditions. In *AIP Conf. Proc.* **1430** (2012) 142, doi: 10.1063/1.4716224.

Chapter 3

Numerical Modeling of Ultrasonic Wave Propagation in Composites

Sourav Banerjee* and Sajan Shrestha[†]

Assistant Professor
Director, Integrated Material Assessment and Predictive Simulation (i-MAPS)
Laboratory Department of Mechanical Engineering
University of South Carolina, 300 Main Street, Columbia, SC 29208, USA
* banerjes@cec.sc.edu
[†] sajan@email.sc.edu

This chapter is designed to provide a brief introductory understanding of wave propagation in composite materials. Waves in composite materials that are considered anisotropic are far more complicated than the waves in isotropic materials. In this chapter, we will cover the fundamentals of wave propagation and will sequentially describe the inevitable mathematical treatments that are required to simulate the elastodynamics in anisotropic media. We will present the understanding of the phenomena that are remarkably different in anisotropic than their counterparts in isotropic materials. Wavefield modeling in anisotropic media requires extensive understanding of the physics and far more computational resources. This chapter will serve as a seedling gesture to the great endeavor that has been currently undertaken by many researchers throughout the world. This chapter presents the summary of the new advancements of the knowledge in the field of modeling ultrasonic waves and presents the current challenges, technical obstacles and the possible direction of the future research.

1. Introduction

In this section, we first ask a fundamental question. Why do we need to understand wave propagation in the engineered materials? Because the primary necessity comes from the computational non-destructive evaluation (Comp-NDE), which requires simulating the elastodynamic phenomena in composite materials,[1] we call virtual NDE. The NDE[2] and the structural health monitoring (SHM)[3] of composites are quite valuable for composite certification process. NDE/SHM are devised to diagnose the health of materials. They provide an understanding of the material state, with which the remaining useful life of the structure can be estimated. Different industries use different NDE techniques. However, the ultrasonic[4] NDE is the most popular due to its capability to see the materials from the inside. SHM has been realized as an online NDE method where traditional handheld ultrasonic transducers can be replaced by the surface-mounted or embedded piezoelectric transducers capable of electromechanical transduction.[5] Irrespective of their mode of energy transduction, ultrasonic waves are introduced inside the material as a probing energy and the propagated energy is sensed from a distant location to diagnose the material in-between. Thus, here onwards, in this chapter, both the traditional NDE and the SHM mode of inspection are included in the term NDE.

By now, it is realized that the quantifiable material state or the damage state in the materials can be derived from the ultrasonic NDE experiments. However, material degradation, complicated damage inside the material like hidden delamination, kissing bonds and trapped cracks are not understood easily or rather not always visualized from the ultrasonic signals. Hence, using the traditional method of NDE interpretation, such damages go unnoticed or undetected. Understanding the ultrasonic wave signals affected by damages is a daunting task and could only be realized if the physics of wave[6] propagation in composites is correctly understood and further simulated numerically. Waves in anisotropic composites with complicated geometries are difficult to understand[7–9] by the traditional ray tracing methods. Similarly, to understand the ultrasonic wave interactions in the composites with various types of possible damages[8,10–13] and material degradation, it is neither feasible nor cost-effective to perform all the possible experiments with a large number of varying damage scenarios and geometries. Hence, now the question is, what should be the solution?

The idea for a most cost-effective solution is a reliable simulation environment that could virtually perform the NDE experiments. These virtual experiments would generate plethora (library) of information from the multiple wave–damage interaction scenarios. Further many such wave–damage interaction scenarios could be used to interpret the quantified fluctuations in the signal. We could learn from such virtual signals to reliably interpret the real-life damage scenarios (e.g., extent of internal damage, degree of material degradation or quantify the material properties[4,10,14–20] etc.) from NDE experiments. This is our fundamental motivation for the numerical modeling of ultrasonic wave propagation. But, the question is, are we there yet to certify the composite virtually through a few experimental signals? No, not yet. Extensive research will be required to push the limit of the computational capability by selecting and optimizing the appropriate methodology, computational resources and/or invent novel methodologies to model the waves in the structures numerically. As mentioned that this chapter is a seedling gesture to set a pathway for wavefield modeling, we await numerous multidisciplinary activities to fulfill the requirements to reach our ultimate goal. Before going into the mathematical details, it is worthwhile to review the work that has already been done in the field of numerical modeling of ultrasonic wave propagation.

2. Background of Wavefield Modeling

During the World War II, to fulfill the demand of quality assurance of the mechanical systems, machine components and the defense equipment, NDE became inevitable. However, after the war, lack of business cases in the defense sectors pushed the majority of the NDE industries to focus on the consumer products due to their enormous industrial growth, while quality assurance of the aerospace systems became the new agenda for government agencies. During the post-world-war era, the ultrasonic NDE pushed their limit to the zenith by designing novel configurations of the ultrasonic transducers for various different applications. Hence, for accurate inspection process, effective design of the transducers and the interpretation of ultrasonic signals required further study to understand the ultrasonic wavefield generated in front of the transducers. The requirement of this understanding was different from the understanding of vibration and wave propagation in the wires, rods, plates and various simpler structural components that were extensively studied even before World War II. In contrast to the plane wave assumption and the plane wave interaction with the structural components, industries had the problem for the first time, where understanding of the wave interaction with finite dimension transducers were required. This requirement leads to mathematical modeling of the ultrasonic wave using Rayleigh–Sommerfeld integral technique or simply called the ray tracing method. Later, this was termed as dynamic ray tracing technique, required the solution of the Eikonal equation,[21] and was used to model the ultrasonic field generated by the ultrasonic transducer. Dynamic ray tracing is an improvement over the previously proposed models with paraxial approximation for the plane wave incidents. It can be realized that in real-life experiments using transducers, the incident wavefronts are neither plane nor spherical. Hence, although we had a solution but was not accurate. To enhance accuracy, researchers used various different computation methods keeping the pace with the technological progress in last 50 years. Some of the popular methods used to simulate the virtual NDE experiments were finite element method (FEM),[22] boundary element method (BEM),[23,24] indirect boundary integral equation (IBIE),[25-28] multi-Gaussian beam model (M-GBM),[7,29,30] charge simulation technique (CST),[31] multiple multi-pole method (MMP),[32-34] spectral element method (SEM),[35] elastodynamic finite integration technique (EFIT),[36] distributed point source method (DPSM),[37-42] Gaussian distributed point source method (G-DPSM),[43] etc. FEM is the most popular method, because there are many commercial packages available to perform these simulations. Unfortunately, spurious reflection at the element boundaries makes the FEM simulations incorrect. All competitive methods have their own pros and cons. Similarly, with the advantages and disadvantages, in this chapter, we will briefly discuss the DPSM method for virtual simulation of the NDE experiments. However, before we go in to the further details of the DPSM, as pre-requisite we should first briefly discuss the wave phenomena in anisotropic material.

3. Waves in Anisotropic Media

Waves in anisotropic media[6] does not propagate with spherical wavefront as it does in isotropic materials. As composites can be anisotropic in their most complex form, we will discuss the waves in fully anisotropic material in a general sense. Say, one has an ultrasonic transducer attached to an anisotropic media emitting the ultrasonic field in the direction of $\mathbf{k}$ vector (Fig. 1(a)). Being inside an isotropic material the wave energy could have gone along the direction of the $\mathbf{k}$ vector. However, being inside an anisotropic media, wave did not take the route $\mathbf{k}$ but created a new wavefront for the wave energy to be propagated. It can be seen that the velocity of the wave energy C_E is at an angle φ with the wavevector $\mathbf{k}$, i.e., along a different direction than the phase velocity direction along C_p. Please note that the plane of energy propagation may not be on the same plane as the incident wave. This means that the $\mathbf{N}$ vector may not be on the $x_1 - x_2$ plane, it can be along any direction in the 3D co-ordinate system based on the material properties and the direction of the $\mathbf{k}$ vector. Additionally,

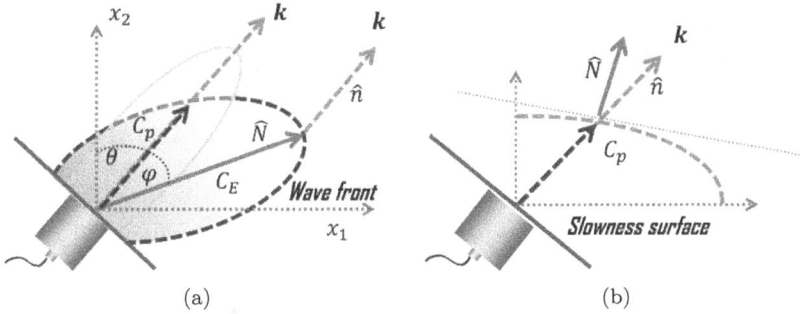

Figure 1. An artistic view showing the schematics of the wavefield inside a bulk anisotropic media (a) showing the wavefront and (b) showing the slowness surface.

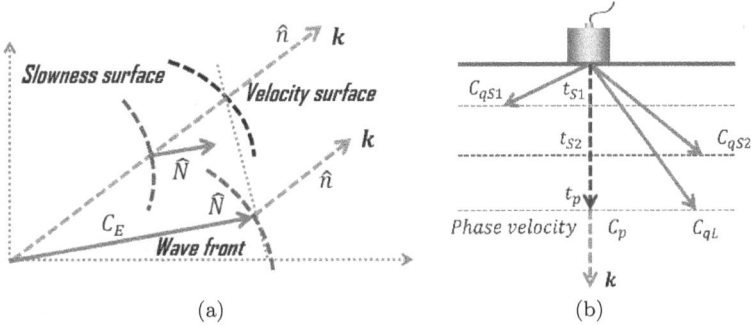

Figure 2. (a) Schematics of the wave surfaces and their relation; (b) relation of the phase velocity with the fundamental wave modes in anisotropic media.

Fig. 1(b) shows the velocity slowness surface, which represents the surface obtained by plotting the inverse of the velocity along the direction of the wave propagation, i.e., $\mathbf{k}$ vector. Normal to that slowness surface conversely represents the direction of the wave energy propagation. It is clear that in the anisotropic material, the wavefront surfaces and the slowness surfaces are not concentric as it is the case for the isotropic materials.

To represents the relationship between these surfaces, we can direct our attention to Fig. 2(a), where three different but very important possible intersecting surfaces are shown. Velocity surface is the surface constructed with the values of the modal wave velocities along the direction of the wavevector $\mathbf{k}$. But that does not mean that wave mode is propagated along the same direction with that same velocity. Now by taking the inverse of the wave velocity magnitudes and if plotted along the same direction of wavevector $\mathbf{k}$, another similar surface emerged, which is called the slowness surface. Interestingly normal to that slowness surface represents the direction of the energy propagation. Next, if the equal energy lines are plotted in the media, one will get the wavefronts. As discussed in Fig. 1(a) that the wavefront is propagated along the direction of the energy propagation, actually the normal to the slowness surface represents the direction of the wavefront as shown in Fig. 2(a). All these phenomena are explainable through the mathematical treatments later discussed in this chapter. They are briefly discussed herein.

For example, when the wave energy is introduced inside the anisotropic media, the wave breaks down in to the three fundamental modes,[6,44] quasi-longitudinal (qL), quasi-shear 1 or quasi-slow shear (qSS) and quasi-shear 2 or quasi-fast shear (qFS). While the wavevector ($\mathbf{k}$) represents the intended direction of the energy propagation of the wave in an anisotropic media, the wave energy propagates in a different direction due to the above-mentioned three wave modes with their respective wave velocities propagated along three different directions. However, projection of the wave modes along the direction of the $\mathbf{k}$ vector will contribute to the phase velocity of the wave packet along the $\mathbf{k}$ (Fig. 2(b)) direction.

Hence, by now, we have realized that it is just not difficult but impossible to track these wave traces using the conventional ray tracing methods in the composites. Indeed, we need a sophisticated mathematical method for modeling the wave in anisotropic media where all the above-mentioned situations can be incorporated by virtue of their mathematical formulations, automatically.

3.1. *Visualizing the Modal Wave Velocities*

Here, we briefly discuss the physics of wave propagation in anisotropic media. It is assumed that reader is familiar with the index notations and also familiar with the basics of continuum mechanics. Further we will skip some obvious mathematical steps that could be understood by reader's discretion. Standard continuum mechanics symbols are used to represent the variables. The fundamental elastodynamic equation (Eq. (1)) in anisotropic material[6] can be viz.

$$\sigma_{ij,j} + f_i = \rho \ddot{u}_i, \qquad (1)$$

where derivative of the stresses are related to the body force and the force due to the dynamic motions, following the Newton's second law. The stress is related to the strain in a linear elastic media[44] through their respective constitutive matrix (Eq. (2)), assuming the material is linear during the NDE inspection due to the very short exposure time of the wave compared to the loading history of the material.

$$\sigma_{ij} = C_{ijml}\varepsilon_{ml}. \tag{2}$$

Using the geometrically linear strain–displacement relation, the elastodynamic equation takes the form in term of displacement[45] as shown in Eq. (3).

$$C_{ijml}\frac{\partial^2 u_m}{\partial x_j \partial x_l} + f_i = \rho \ddot{u}_i. \tag{3}$$

Now we would need to solve this complicated equation to realize the facts about the anisotropic waves. In NDE experiments, we use a transducer with a central frequency of actuation, which means that the transducer is able to generate the maximum amplitude of displacement at that frequency. However, if not impossible, it is extremely difficult to make a monochromatic ultrasonic transducer. So, when the transducer is actuated, there will be other proximal frequencies generating displacements with gradually lower amplitudes. Assuming linear materials where superposition theorem is valid in the Fourier domain and the reciprocity theorem is valid, we will proceed to solve the above equation using a monochromatic harmonic displacement function and such solutions can be commutable to the other neighboring frequencies generated by the transducer. Hence, let's assume a monochromatic displacement function

$$u_m = Ag_m e^{i(\mathbf{k}.\mathbf{x} - \omega t)}, \tag{4}$$

where A is the scalar amplitude of the wave, g_m is the polarization direction, ω is the monochromatic wave frequency, $\mathbf{k}$ is the wavevector, $\mathbf{x}$ is the position vector where, $\mathbf{k}.\mathbf{x}$ represents the dot product between $\mathbf{k}$ and $\mathbf{x}$ represents the phase component of the wave. After substituting Eq. (4) into Eq. (3) and following few mathematical manipulation,[45] we get the following equation Eq. (5):

$$[C_{ijml}k_j k_l - \rho\delta_{im}\omega^2]Ag_m = -f_i. \tag{5}$$

To solve for the eigenvalues and the eigenvectors of a system, we need to consider the system unperturbed by the forcing functions or alternatively to solve the homogeneous equation. Hence, for the time being, we will set the body force to zero and the non-trivial solution of the equation can be written as

$$[C_{ijml}n_j n_l - \rho \delta_{im}c^2]g_m = 0, \tag{6}$$

where $c^2 = \omega^2/k^2$ is the phase velocity of the wave along the direction of $\mathbf{k}$ vector, n_j are the direction cosine of the wave propagation direction, i.e., again along the $\mathbf{k}$ vector. The solution of this equation give us the wave modes that are propagating in the material with material constants C_{ijml}. Equation (6) is the well-known Christoffel's equation.[6] The solution of this equation will provide three eigen modes with wave velocities C_{qL}, C_{qFS} and C_{qSS}, respectively, along the directions found from the eigenvectors. Figure 3 shows the understanding of the solution of Eq. (6) in a pictorial form. The solution is performed for a specific direction of wave propagation along the $\mathbf{k}$ vector with direction normal $\mathbf{n}$. In a 3D co-ordinate system, the wave velocities are at the direction of their eigenvectors $\boldsymbol{\varphi_L}$, $\boldsymbol{\varphi_{FS}}$, $\boldsymbol{\varphi_{SS}}$, respectively. Immediately now, it can be realized that the projection of the phase velocities of the wave modes on the $\mathbf{k}$ vector gives the phase velocity of the wave along $\mathbf{k}$. However, direction of the wave energy will be at the resultant direction with their group velocities. Hence, this section has equipped us with the tool to find the modal wave velocities due to any direction of wave actuation, i.e., just by changing the $\mathbf{n}$ vector pointing to any direction in the 3D coordinate system.

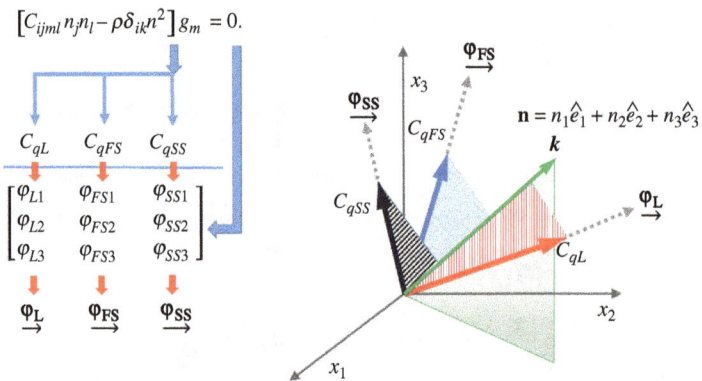

Figure 3.   Pictorial representation of the Christoffel's solution where the modal phase wave velocities are the eigenvalues and the direction of those wave modes are the eigenvectors.

3.2. *Solution of Elastodynamic Green's Function*

Now we have a clear understanding of the wave modes in bulk anisotropic media. Further to tailor our discussion to the simulation of a virtual NDE experiment using DPSM, we need to find the elastodynamic Green's function in the anisotropic media,[9,46–50] where the fundamental solution of Eq. (6) will be crucial. Recollecting Eq. (5), this time we cannot make the body force zero to make the problem homogeneous. Instead, we will introduce a pulse force in the media due to a point source, which by definition will give the solution of the elastodynamic Green's function. Based on this development, we modify Eq. (5) where $G_{mp}(x_n, t)$ is the time domain Green's function at x_n due to a point source actuating along the pth direction with force amplitude F_p.

$$\left[C_{ijml} \frac{\partial^2}{\partial x_j \partial x_l} - \rho \delta_{im} \frac{\partial^2}{\partial t^2} \right] G_{mp}(x_n, t) = -\delta_{ip}\delta(x_n)\delta(t)F_p. \tag{7}$$

Next for an unit amplitude of actuation, by transforming Eq. (7) to the Fourier domain ($\mathcal{F}$) for both spatial and temporal variables, we get the governing elastodynamic equation to find the Green's function at a point x_n in the frequency–wavenumber domain as follows:

$$[C_{ijml}k_j k_l - \rho\omega^2\delta_{im}]\tilde{G}_{mp}(k_n, \omega) = -\delta_{ip}\frac{1}{(2\pi)^3}. \tag{8}$$

Please note that we used $\mathcal{F}[\delta(x_n)] = 1/(2\pi)^3$. Considering an operator in the $\mathcal{L}^2$ Hilbert space the above equation can be written as

$$\mathbf{L}_{im}\tilde{G}_{mp}(k_n, \omega) = \delta_{ip}\frac{1}{(2\pi)^3}, \tag{9}$$

where

$$\tilde{G}_{mp}(k_n, \omega) = \frac{1}{(2\pi)^4} \iiint \int_{-\infty}^{\infty} \int_{-\infty}^{\infty} G_{mp}(x_n, t)e^{i(k_l x_l - \omega t)}dx^3 dt, \tag{10}$$

$$\mathbf{L}_{im}(k_n, \omega) = [C_{ijml}k_j k_l - \rho\omega^2\delta_{ij}]. \tag{11}$$

As we have the basic dynamic equation in the frequency–wavenumber domain, we could further transform back the equation to the frequency–space domain and the frequency domain Green's function in terms of the operator and can be written as

$$g_{mp}(x_n, \omega) = \frac{1}{(2\pi)^3} \iiint_{-\infty}^{\infty} [\mathbf{L}_{mp}(k_n, \omega)^{-1}] e^{-ik_l x_l} dk^3. \tag{12}$$

The above equation is very interesting in itself, which says that the Green's function at any point x_n in space will be the integral of all the possible wave numbers, which means that the Green's function is the superposition of the influences from waves propagating in all the possible directions.

3.3. *Wave Modes in all Possible Wave Directions in 3D*

Now before going in to further discussion of Green's function, we will explore the solution obtained from Section 3.1. The reason for referring back to Section 3.1 is the term $\mathbf{L}_{im}(k_n, \omega)$, which is inherently associated with the solution of Eq. (6). After numerically solving Eq. (6), we are able to obtain the eigenvalues and eigenvectors for each direction of propagation of the waves. Let us assume an anisotropic material GaAs with the following material constants

$$\begin{bmatrix} C_{11} & C_{12} & C_{13} & C_{14} & C_{15} & C_{16} \\ C_{21} & C_{22} & C_{23} & C_{24} & C_{25} & C_{26} \\ C_{31} & C_{32} & C_{33} & C_{34} & C_{35} & C_{36} \\ C_{41} & C_{42} & C_{43} & C_{44} & C_{45} & C_{46} \\ C_{51} & C_{52} & C_{53} & C_{54} & C_{55} & C_{56} \\ C_{61} & C_{62} & C_{63} & C_{64} & C_{65} & C_{66} \end{bmatrix}$$

$$= \begin{bmatrix} 72 & 10.3 & 10.3 & 0.00 & 0.00 & 0.00 \\ 10.3 & 78.3 & 24.5 & 0.00 & 0.00 & 0.00 \\ 10.3 & 24.5 & 78.3 & 0.00 & 0.00 & 0.00 \\ 0.00 & 0.00 & 0.00 & 5.00 & 0.00 & 0.00 \\ 0.00 & 0.00 & 0.00 & 0.00 & 175.0 & 0.00 \\ 0.00 & 0.00 & 0.00 & 0.00 & 0.00 & 175.0 \end{bmatrix} \text{GPa.} \tag{13}$$

Figure 4 shows the three-wave velocity slowness, which is inverse of the wave velocities in all possible directions of the wave propagation by discretizing a sphere. Figures 4(a)–4(c) shows the wave velocity slowness of the qL, qSS (qS1), qFS (qS2) modes, respectively. Similarly, Fig. 5 shows the wave velocity slowness of the longitudinal and two shear wave modes in

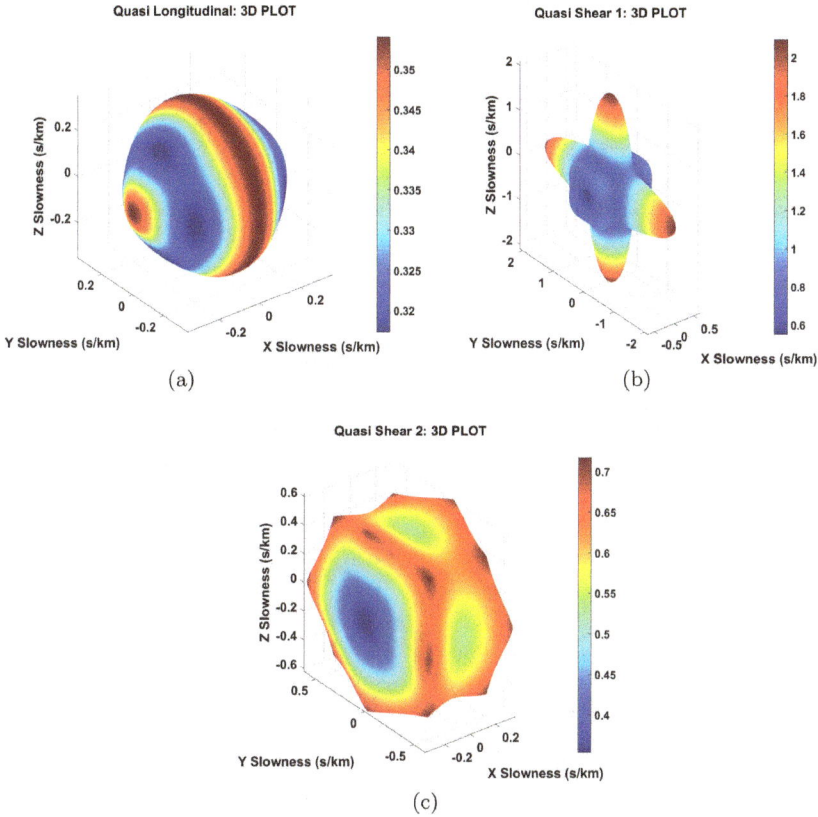

Figure 4. 3D slowness plot for GaAs in (a) quasi-longitudinal mode, (b) quasi-slow shear (qS1) mode and (c) quasi-fast shear (qS2) mode.

aluminum, which is an isotropic material. The slowness surfaces, which are spherical in isotropic material are non-spherical in anisotropic material.

Similarly, we can take different anisotropic materials with representative material properties to find the similar wave modes in 3D considering all possible wave directions. For example, three different materials are considered: (a) orthotropic, transversely isotropic with material properties $C_{11} = 143.8$ GPa, $C_{12} = C_{13} = 6.2$ GPa, $C_{22} = 13.3$ GPa, $C_{23} = 6.5$ GPa, $C_{33} = 13.3$ GPa, $C_{44} = 3.4$ GPa, $C_{55} = C_{66} = 5.7$ GPa; (b) fully orthotropic material with material properties $C_{11} = 70$ GPa, $C_{12} = 23.9$ GPa, $C_{13} = 6.2$ GPa, $C_{22} = 33$ GPa, $C_{23} = 6.8$ GPa, $C_{33} = 14.7$ GPa, $C_{44} = 4.2$ GPa, $C_{55} = 4.7$ GPa, $C_{66} = 21.9$ GPa; (c) monoclinic material with material properties $C_{11} = 102.6$ GPa, $C_{12} = 24.1$ GPa,

Quasi Longitudinal: 3D PLOT

(a)

Quasi Shear 1: 3D PLOT

(b)

Quasi Shear 2: 3D PLOT

(c)

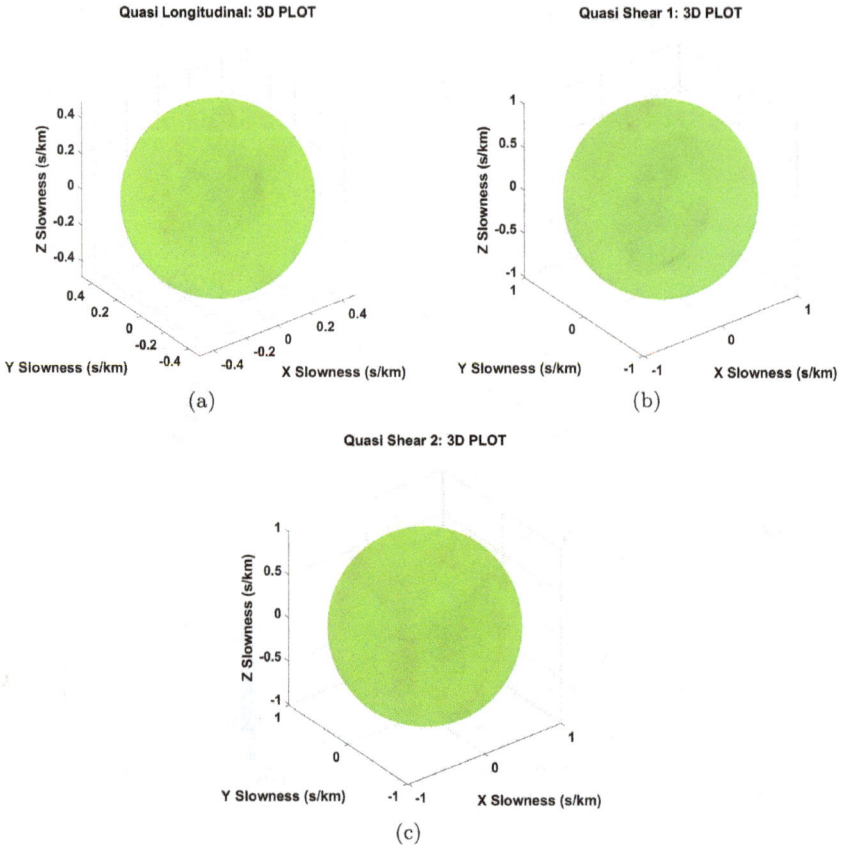

Figure 5.   3D slowness plot for aluminum: (a) longitudinal mode; (b) shear (qS1) mode in isotropic; and (c) shear (qS2) mode in isotropic.

$C_{13} = 6.3$ GPa, $C_{16} = 40$ GPa, $C_{22} = 18.7$ GPa, $C_{23} = 6.4$ GPa, $C_{26} = 10$ GPa, $C_{33} = 13.3$ GPa, $C_{36} = -0.1$ GPa, $C_{44} = 3.8$ GPa, $C_{45} = 0.9$ GPa, $C_{55} = 5.3$ GPa, $C_{66} = 23.6$ GPa are considered. Figures 6–8 respectively shows the wave velocity slowness pattern in the above-mentioned materials, respectively.

Please note that the slowness profile is circular in y–z plane for the transversely isotropic material (Fig. 6(a)), which is expected. The same circular profile is prevailed in monoclinic material with a skewed axis of symmetry.

The calculation of the wave velocity slowness in three dimension are crucial for the next step to calculate the 3D Green's function in the materials.

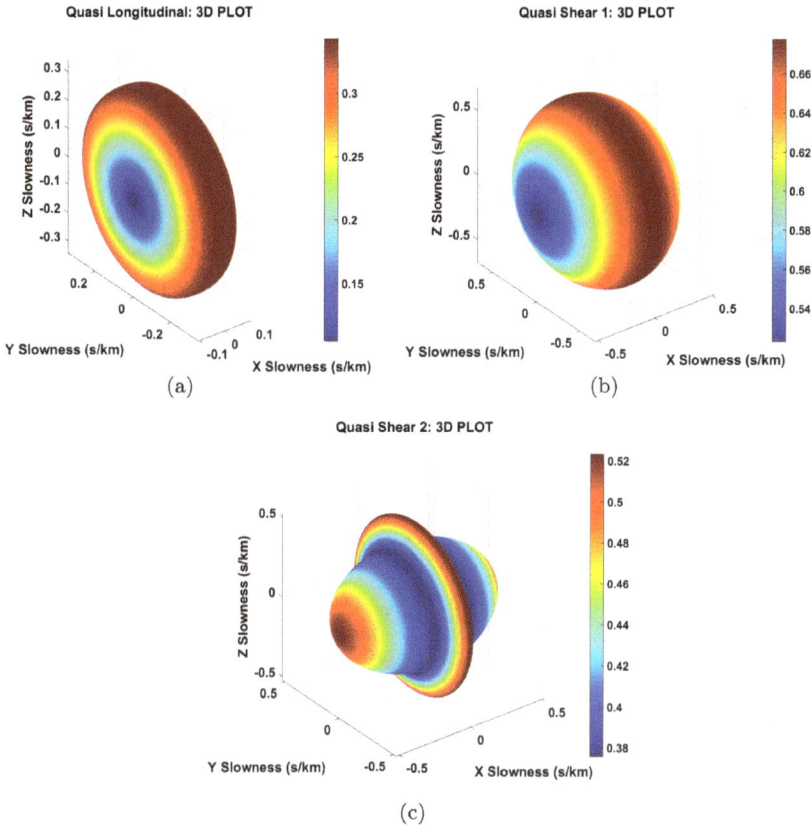

Figure 6. 3D slowness plot for Transversely Isotropic, orthotropic composite material (a) quasi-longitudinal mode; (b) quasi-slow shear (qS1) mode; (c) quasi-fast shear (qS2) mode.

3.4. *Exact Mathematical Expression for the Green's Function*

From Section 3.2 we have understood the mathematical meaning of the Green's function. However, to compute the Green's function in the frequency domain, we need a more tractable form. This section is dedicated to explaining the steps to find the mathematical expression of Green's function that can be computed numerically. In this section, we also explain the meaning of each term that goes in to the computation of the Green's function. Now we take the same equation (Eq. (7)) but treat with a different process. Instead of applying the Fourier transform for the space and time

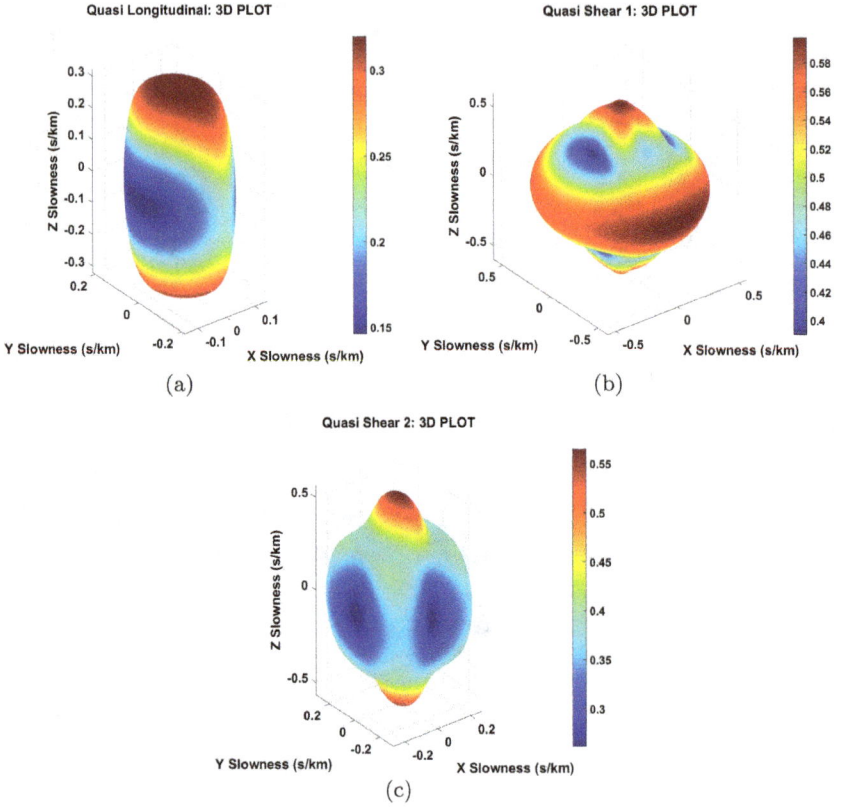

Figure 7. 3D slowness plot for fully orthotropic composite material (a) quasi-longitudinal mode; (b) quasi-slow shear (qS1) mode; and (c) quasi-fast shear (qS2) mode.

both we just transform the equation to the frequency domain with unit amplitude force:

$$\left[C_{ijml} \frac{\partial^2}{\partial x_j \partial x_l} + \rho \omega^2 \delta_{im} \right] g_{mp}(x_n, \omega) = -\delta_{ip} \delta(x_n). \tag{14}$$

Note that now we have the frequency domain Green's function in Eq. (14). However, the solution of these partial differential equations (Eq. (14)) will be very complicated and thus we need to find a possible method where these partial differential equations can be transformed into a set of ordinary differential equations, which can be solved easily.

As an alternative approach to the Fourier transform method, it is proposed to use an elegant transformation called Radon's transform, which

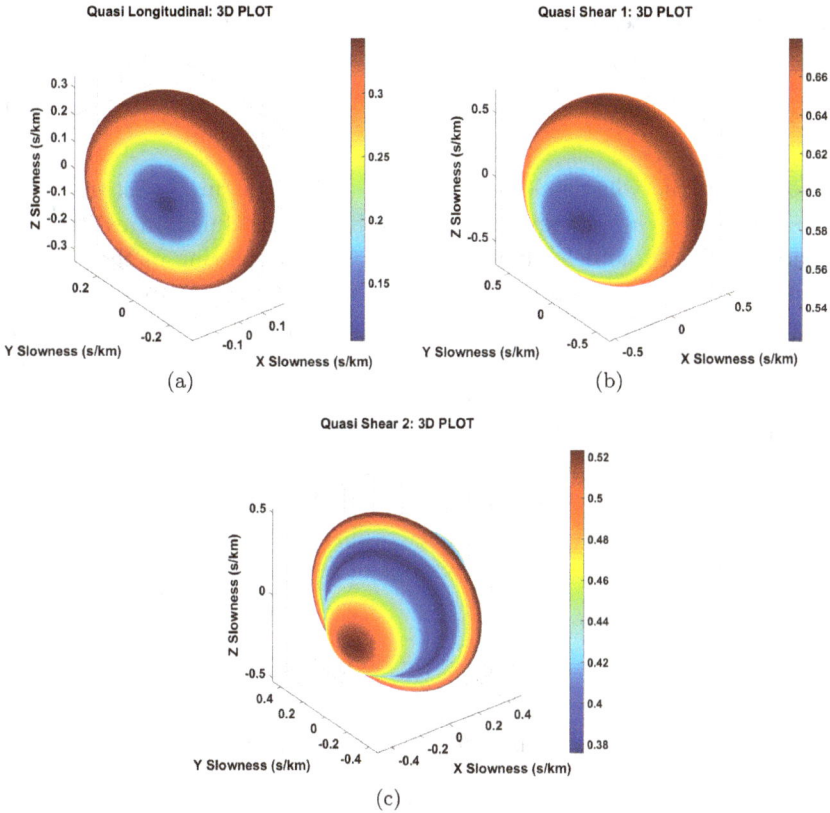

Figure 8. 3D slowness plot for monoclinic composite material: (a) quasi-longitudinal mode; (b) quasi-slow shear (qS1) mode; (c) quasi-fast shear (qS2) mode.

is currently very popular for the computed tomography (CT) and 3D image reconstruction from a cluster of 2D projection images taken on different planes. Hence, if the images on the 2D planes can be transformed to a 3D image, then it is also possible for a 3D function to be transformed to multiple functions projected on 2D planes and parameterized by their orientation of the planes. When we do the CT, we actually perform an inverse Radon transform. But a forward Radon transform is transferring a 3D function to several 2D functions, parameterized by the definition of the 2D planes. Let us discuss the basics of Radon transform mathematically. If in an Euclidean space $\Re^3$ with x_1, x_2 and x_3 are the co-ordinates of a point $(\mathbf{x})$ in space, Radon transform[45] of a function $g(\mathbf{x})$, which is defined and absolutely integrable over all space, can be written as

$$\tilde{g}(\vec{\mu}, h) = R\{g(\vec{x})\} = \int g(\mathbf{x})\delta(h - \vec{\mu} \cdot \vec{x}) d\Omega, \tag{15}$$

where $d\Omega$ is the volume element and Eq. ((15)) is a surface integral over the planes $h = \mu_k x_k$, only where the values of the function exists after operating the Dirac delta function. The physical meaning of the integral could be briefly described in 2D and 3D using Fig. 9. There are infinite numbers of lines in 2D and planes in 3D at different distances h from the origin with their local orientation of the lines (2D) or planes (3D), designated by their unit normal $\hat{\mu}$. The Dirac δ function in Eq. (15) signifies that the values exists only on the plane $\hat{\mu}, h$ and elsewhere the integral is zero.

Radon transform has many elegant basic features described below which we applied to obtain the coupled ordinary differential equations from Eq. (14). Radon transform of derivatives of a function can be written as[9]

$$R\left\{\left(\frac{\partial^2 g}{\partial x_i \partial x_j}\right)\right\} = \mu_i \mu_j \frac{\partial \tilde{g}(h, \mu)}{\partial h^2}. \tag{16}$$

And the inverse Radon transform of the above equation can be written as

$$g(\mathbf{x}, \omega) = -\frac{1}{8\pi^2} \iiint_{|\boldsymbol{\mu}|=1}^{\text{sphere}} \left. \frac{\partial^2 \tilde{g}(h, \boldsymbol{\mu})}{\partial h^2} \right|_{h=\boldsymbol{\mu} \cdot \mathbf{x}} d\Omega(\boldsymbol{\mu}) \tag{17}$$

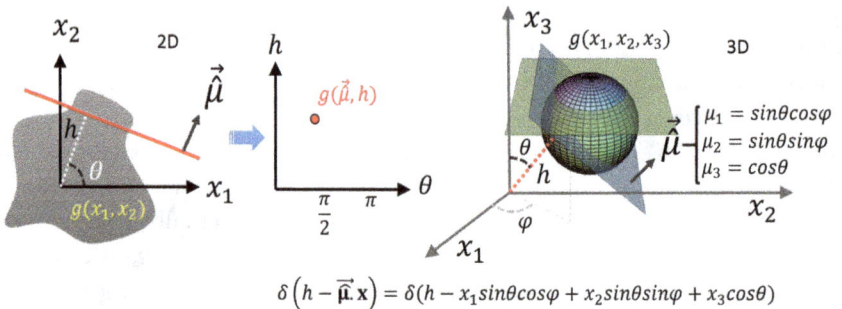

$$\delta\left(h - \vec{\hat{\mu}} \mathbf{x}\right) = \delta(h - x_1 sin\theta cos\varphi + x_2 sin\theta sin\varphi + x_3 cos\theta)$$

Figure 9.   A schematic diagram to visualize the Radon transform in 2D and 3D. The figure shows that the integral is performed along a line at different distances h along $\vec{\mu}$ in 2D and over a plane at different distances h along $\vec{\mu}$ in 3D.

We will use these identities in our foregoing calculations. After applying the Radon transform on Eq. (14), we get coupled ordinary differential equations

$$\Gamma_{im}\frac{\mathrm{d}^2\tilde{g}_{mp}}{\mathrm{d}h^2} + \omega^2\delta_{im}\tilde{g}_{mp} = -\delta_{ip}\frac{1}{\rho}\delta(h), \tag{18}$$

$$\Gamma_{im} = \frac{C_{ijml}\mu_j\mu_l}{\rho}; \quad B_{mj}B_{jp} = \omega^2\Gamma_{im}^{-1}. \tag{19}$$

Fortunately, while decoupling the above set of coupled ordinary differential equations, we find that the eigenvalue solution that is required have already been solved using Eq. (6) and the solutions are depicted in Fig. 3. The wave propagation direction pointed by the direction cosine n_i in Eq. (6) are equivalent to the direction cosine of the plane used for integration in Radon transform in Eq. (16). The solution[50] of Eq. (18) can be viz.

$$\tilde{g}_{mp} = \frac{i}{\omega^2\rho}f_pB_{jp}e^{ihB_{mj}}; \quad h > 0. \tag{20}$$

Further applying the spectral theorem, we could write an analytic function **L**, which is a function of the Γ_{im} matrix in the following form:

$$\mathbf{L}(\Gamma_{im}) = \sum_{z=1}^{n}\mathbf{L}(\gamma_z)(P_{im})^{(z)}, \tag{21}$$

where z takes the index of eigenvalues to run the sum and n is the maximum number of eigenvalues in the system. In this case, we have three eigenvalues, thus $n = 3$. γ_z is the zth eigenvalue and $(P_{im})^{(z)}$ is the projection matrix of the zth eigen mode. Here, we will draw some correlation between the eigen modes with the wave velocities C_{qL}, C_{qFS}, C_{qSS}, their eigenvectors $\boldsymbol{\varphi_L}$, $\boldsymbol{\varphi_{FS}}$, $\boldsymbol{\varphi_{SS}}$, respectively and the new parameter defined as γ_z and $(P_{im})^{(z)}$. The relation can be explicitly written as follows:

$$\gamma_1 = (C_{qL})^2; \gamma_2 = (C_{qFS})^2; \gamma_3 = (C_{qSS})^2, \tag{21}$$

$$(P_{im})^{(1)} = \varphi_{Li}\varphi_{Lm}; (P_{im})^{(2)} = \varphi_{FSi}\varphi_{FSm}; (P_{im})^{(3)} = \varphi_{SSi}\varphi_{SSm}. \tag{22}$$

Based on this understanding, we further write the spectral expression for the analytic transformed displacement function as

$$\tilde{g}_{mp} = \frac{i}{\omega^2\rho}f_p\sum_{z=1}^{3}\frac{\omega}{\sqrt{\gamma}_z}e^{ih(\frac{\omega}{\sqrt{\gamma}_z})}(P_{mp})^{(z)}. \tag{23}$$

Now applying the inverse Radon transform, we get the displacement Green's function[50] in the frequency domain

$$g_{mp}(x_n, \omega) = \frac{i\omega}{2(2\pi)^2 \rho} \sum_{z=1}^{3} \iint_{\theta=0;\phi=0}^{\theta=\pi/2;\phi=2\pi}$$

$$\times \left[\left(\sqrt{1/\gamma_z} \right)^3 (P_{mp})^{(z)} \exp(i(k_i x_i)^{(z)}) d\Omega(\mu) \right]$$

$$+ \frac{1}{8\pi^2} \int_{|n|=1} \Gamma_{mp}^{-1}(\mu) \delta(\mu.x) d\Omega(\mu). \tag{24}$$

Alternatively, in Ref. [50] $H_z(\mu_i x_i)$ the Heaviside step function is introduced to impose the vanishing integral for $\mu_i x_i < 0$, which means that the integral is valid only when the angle between the wave propagation directorate and the Radon's plane is less than $90°$. We could say that the target–source direction and the ultimate Radon planes will make a hemisphere, inside which the integral (Eq. (24)) is valid.

Here, now with Fourier approach, we will bring the discussion on Eq. (12) and will apply the theorem discussed in Eq. (21). Using the spectral theorem (Eq. (21)), after few mathematical manipulations, we could write the inverse of the Christoffel's operator $\mathbf{L}_{mp}(k_n, \omega)$ as follows:

$$\mathbf{L}_{mp}(k_n, \omega)^{-1} = \sum_{z=1}^{3} \frac{\varphi_{mi}^{(z)} \varphi_{ip}^{(z)}}{\rho \gamma_z |k|^2 - \rho \omega^2} = \sum_{z=1}^{3} \frac{\gamma_z^{-1} (P_{mp})^{(z)}}{\rho(|k|^2 - \omega^2/\gamma_z)}. \tag{25}$$

Substituting Eq. (25) in to Eq. (12), we get an alternative expression for the displacement Green's function, which we could write explicitly in the spherical coordinate system. However, results from Eq. (24) and our new following form must be equal.

$$g_{mp}(x_n, \omega) = \frac{1}{(2\pi)^3} \sum_{z=1}^{3} \iiint_{|r|=1}^{sphere} [(s_z)^2 (P_{mp})^{(z)} \sin\theta d\theta d\phi]$$

$$\times \int_{-\infty}^{\infty} \left[\frac{\exp(-i|k|(x_1 \sin\theta \cos\phi + x_2 \sin\theta \sin\phi + x_2 \cos\theta)}{(|k|^2 - \omega^2(s_z)^2)} \right]$$

$$\times |k|^2 dk \tag{26}$$

The second part of the integral has a pole at $|k|^2 = a^2$, where $a = \omega s_z$. Applying the following identity (Eq. (27), Cauchy's integral formula) and performing few mathematical simplifications,[46] we get the final expression for the displacement Green's functions in Eq. (28).

$$\int_{-\infty}^{\infty} \frac{\exp(iku)}{(k^2 - a^2)} k^2 dk = \pi i a e^{ia|u|} + 2\pi\delta(u),$$ (27)

$$g_{mp}(x_n, \omega) = \frac{i\omega}{2(2\pi)^2 \rho} \sum_{z=1}^{3} \int\!\!\int\!\!\int_{r=0;\,\theta=0;\,\phi=0}^{r=1\theta=\pi;\,\phi=2\pi}$$

$$\times [(s_z)^3 (P_{mp})^{(z)} \exp(i(k_i x_i)^{(z)}) \sin\theta d\theta d\phi]$$

$$+ \frac{1}{2(2\pi)^2 \rho |x|} \int_0^{2\pi} (s_z)^2 (P_{mp})^{(z)} d\phi.$$ (28)

In the following picture, we schematically explain the integral in Eq. (28).

For each source point (red sphere) and target point (black dot with blue ring) combination (blue arrow), the integral in Eq. (28) is performed for all possible wave directions (green arrow). For each possible direction of the green arrow, three wave modes (see Fig. 3) contributes to the calculation of Green's function. The integral is valid only for the grid points on the sphere those are above the mid-plane (gray plane) perpendicular to the blue arrow. This is synonymous to the Heaviside step function in Ref. [50]. It is easily recognized that the normal to the gray plane can be visualized as a radon plane $h = \mu_k x_k$ in Fig. 10 where μ_k is synonymous to n_k. The second

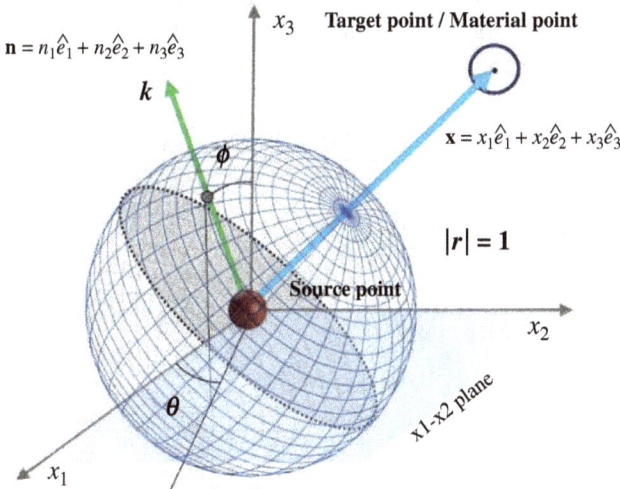

Figure 10.   A schematic diagram to visualize the integral in Eq. (28).

part of the integral in Eq. (28) is performed along the perimeter of the circle on that plane (the mid-plane perpendicular to the blue arrow). Let's change our nomenclature of displacement Green's function for the following mathematical steps. Then we will write $g_{mp} = u_m^p$. Now to calculate the stresses, we will use the constitutive equation for the anisotropic medium given by

$$\sigma_{ik}^p(\mathbf{x}) = C_{ikmj}\varepsilon_{mj}^p(\mathbf{x}), \tag{29}$$

where σ_{ik}^n is the stress tensor at the target point ($\mathbf{x}$) due to the source point with force along the pth direction. The strains at the target point can be calculated using the displacement Green's function obtained from Eq. (28).

$$\varepsilon_{mj}^p = \frac{1}{2}(u_{m,j}^p + u_{j.m}^p). \tag{30}$$

Taking spatial derivative of the displacement Green's functions in Eq. (28) as depicted in Eq. (30) and multiplying the result with the constitutive matrix of the composite material in Eq. (29), we can calculate the stress tensor for a specific direction of point force along p.

4. Numerical Computation of Green's Function

To calculate the anisotropic Green's functions, we will focus our computation on few specific types of materials as we already considered in Section 3: (a) transversely isotropic, (b) fully orthotropic and (c) Monoclinic material. Let us consider two geometric configurations as shown in Fig. 11. Configuration 1, where we present the displacement Green's function

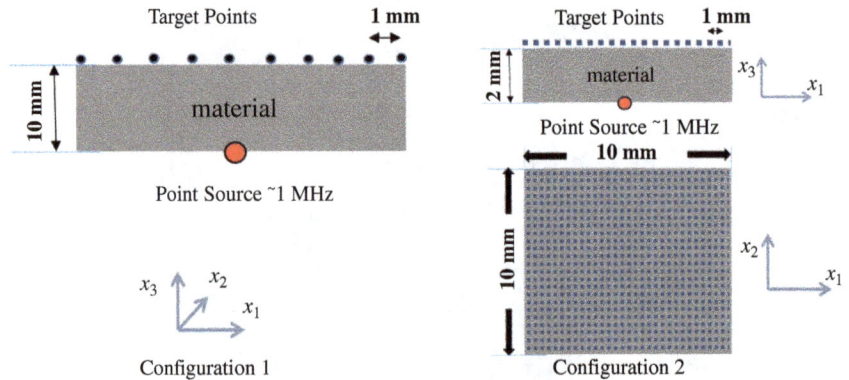

Figure 11.   Schematics of two configurations to calculate the Green's function.

along a straight line inside the materials composed of 10 target points. Configuration 2 where we present the 2D pattern of both the displacement and stress Green's functions on a rectangular domain. Next we present the numerical computation of the displacement Green's function in transversely isotropic material over a plane where 51×51 target points are distributed as shown in Configuration 2 in Fig. 11.

Based on these displacements and stress Green's functions (Figs. 12–17), we enhance our numerical computation and proceed with the calculation of ultrasonic wavefield in an anisotropic solid (we considered a transversely isotropic media) when the ultrasonic bounded beam is

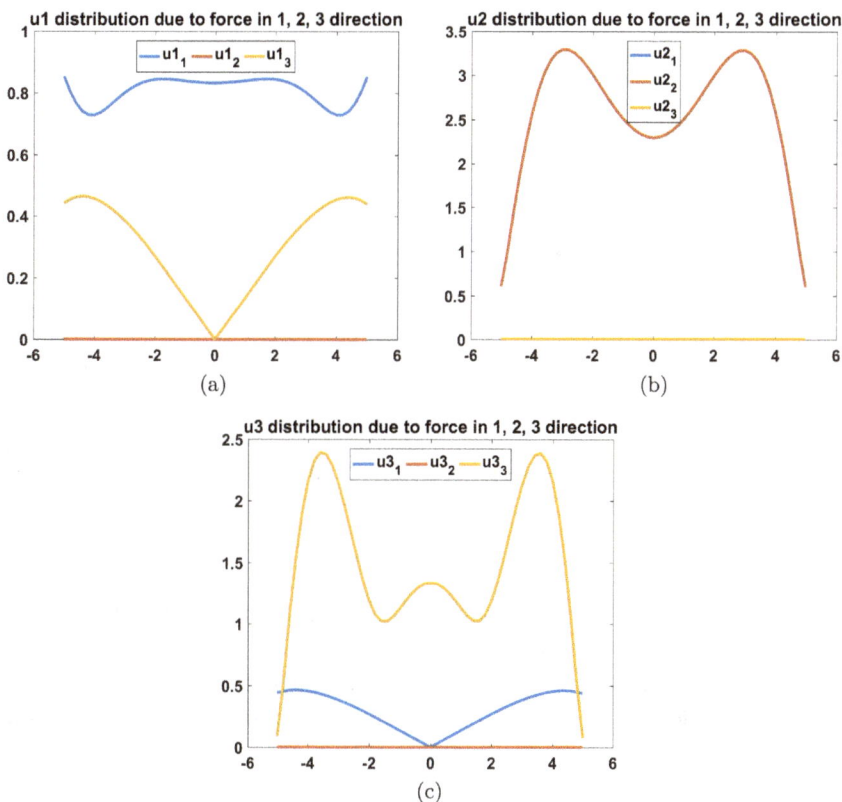

Figure 12. Numerical computation of displacement Green's function (unit mm) due to forces acting along 1, 2 and 3 directions in Configuration 1 when the inside material is transversely isotropic. (a) Displacements along x_1 direction, (b) displacements along x_2 direction and (c) displacements along x_3 direction.

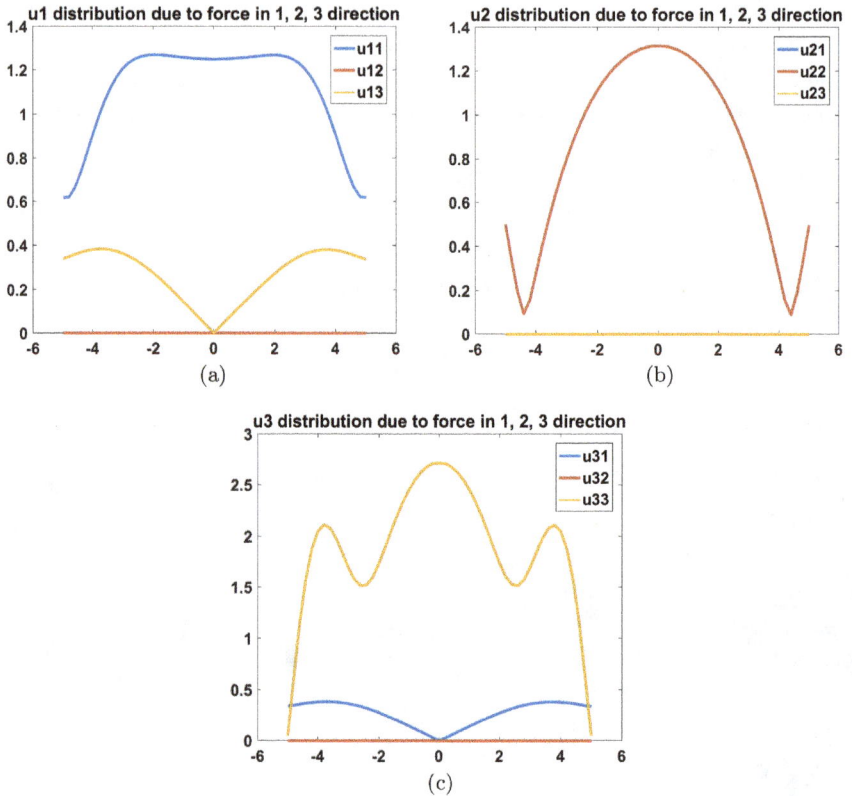

Figure 13. Numerical computation of displacement Green's function (unit mm) due to forces acting along 1, 2 and 3 directions in Configuration 1 when the inside material is fully orthotropic. (a) Displacements along x_1 direction; (b) displacements along x_2 direction; (c) displacements along the x_3 direction due to forces acting along 1, 2 and 3 directions.

generated from an 1 MHz ultrasonic transducer. Here we depict a rather simple problem in Fig. 18(a).

A conventional ultrasonic NDE transducer and an anisotropic half space is considered, which are both immersed in fluid. Following the DPSM process, the active point sources are distributed below the transducer face and two other layers of point sources are distributed on either side of the interface between the two media. The contributions of different point sources are shown by the dotted lines, connecting the relevant point sources and the points of interest (C, D). Then the total ultrasonic field is produced in the fluid and anisotropic media by the superposition of the

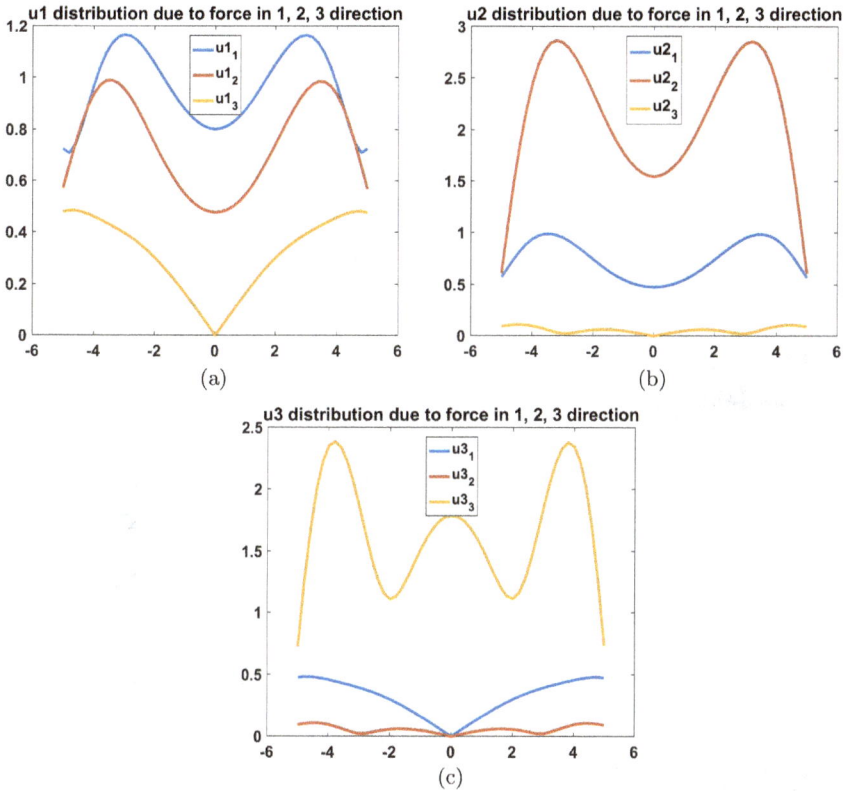

Figure 14. Numerical computation of displacement Green's function (unit mm) due to forces acting along 1, 2 and 3 directions in Configuration 1 when the inside material is monoclinic. (a) Displacements along the x_1 direction; (b) displacements along the x_2 direction; and (c) displacements along the x_3 direction due to forces acting along 1, 2 and 3 directions.

wavefield created by all the point sources; however, the source strengths are unknown. For the current problem in our hand, the length of the interface was taken 10 mm and 55 point sources were distributed on either side of the interface following the wavelength–source diameter rule explicitly described in the previous articles.[38,45] A circular transducer of 2 mm diameter with central frequency ~1 MHz was considered submerged into the water and 100 point sources were distributed, concentrically just below its transmitting face. The distance between the transducer and interface was taken to be 5 mm. The source strengths were $\mathbf{A}_s$, $\mathbf{A}_I$ and $\mathbf{A}_{I^*}$ as shown in Fig. 18(a). To actuate the transducer, unit velocity was prescribed on the transducer face. By using the standard process[38,45] of DPSM, we formulated the

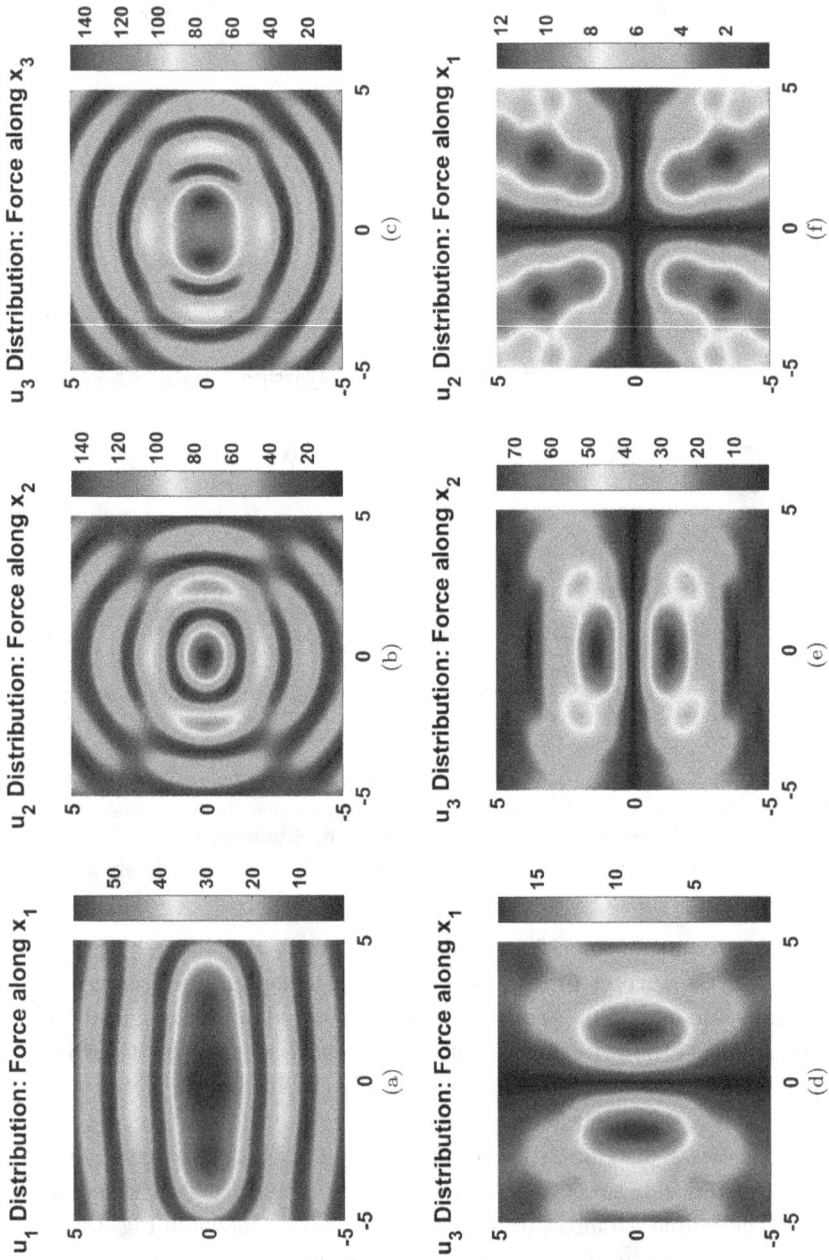

Figure 15. Numerical computation of displacement Green's function (unit mm) in transversely isotropic material: (a) g_{11}; (b) g_{22}; (c) g_{33}; (d) g_{31} or g_{13}; (e) g_{32} or g_{23}; (f) g_{21} or g_{12}.

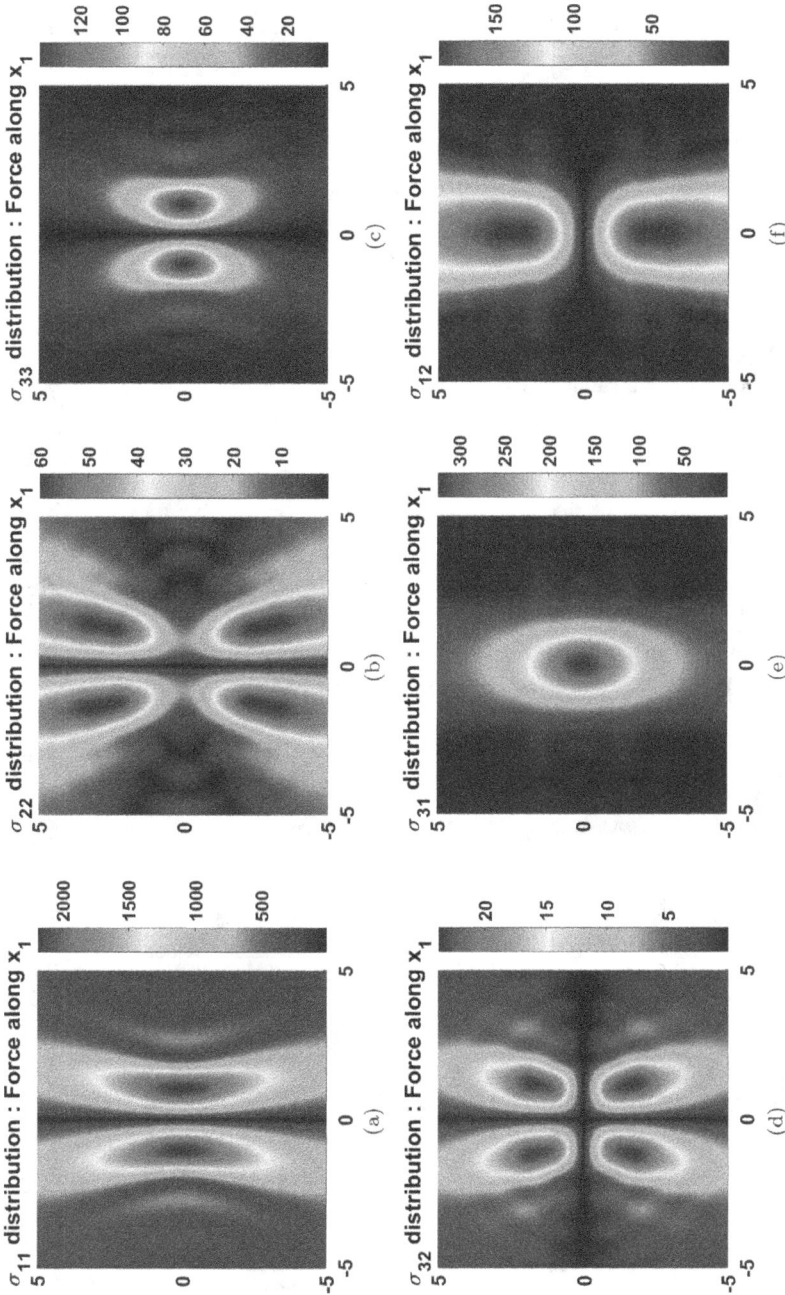

Figure 16. Numerical computation of stress Green's function (unit GPa) in transversely isotropic material due to source actuating along one direction. (a) σ_{11}; (b) σ_{22}; (c) σ_{33}; (d) σ_{32} or σ_{23}; (e) σ_{31} or σ_{13}; (f) σ_{21} or σ_{12}.

S. Banerjee & S. Shrestha

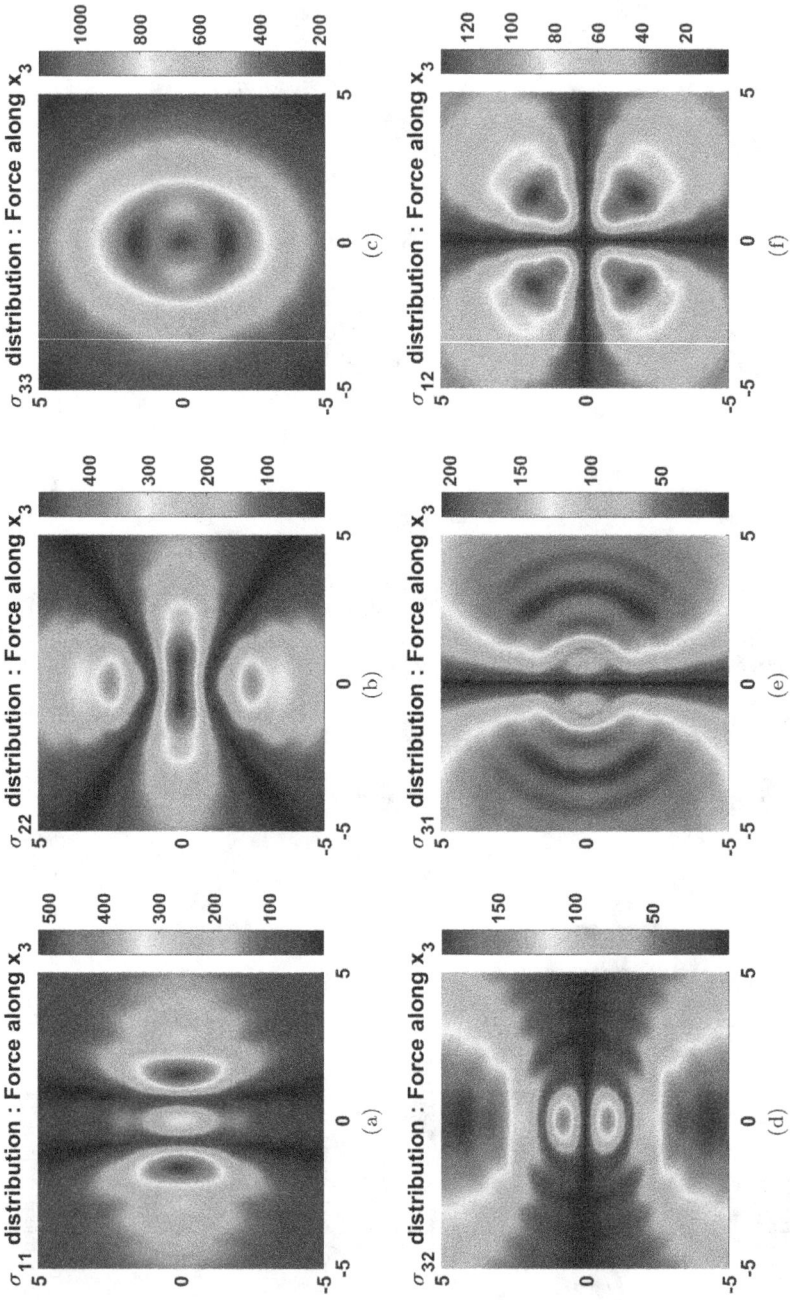

Figure 17. Numerical computation of stress Green's function (unit GPa) in transversely isotropic material due to source actuating along three direction. (a) σ_{11}; (b) σ_{22}; (c) σ_{33}; (d) σ_{32} or σ_{23}; (e) σ_{31} or σ_{13}; (f) σ_{21} or σ_{12}.

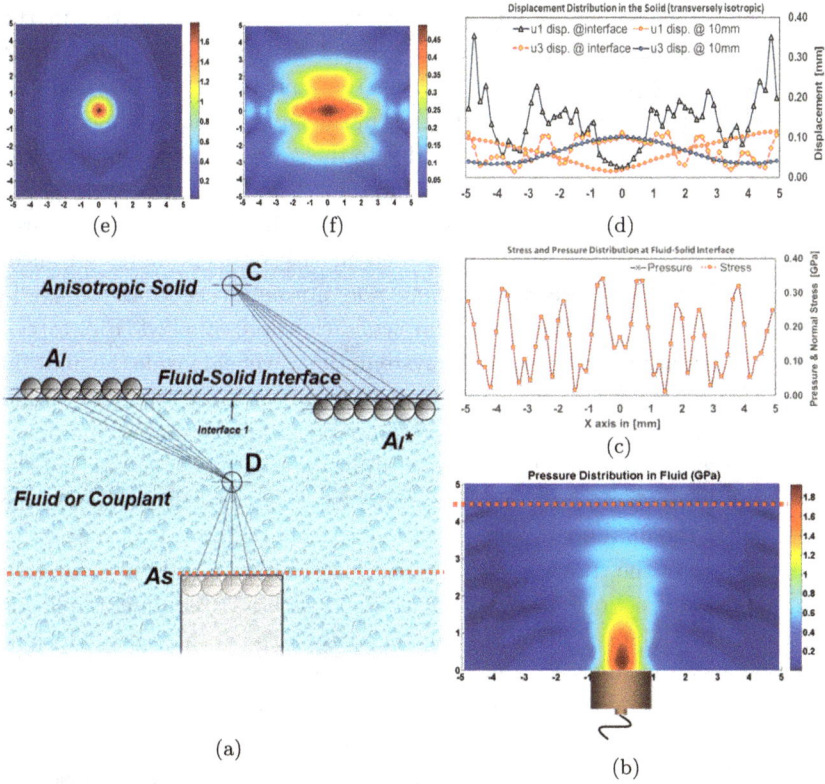

Figure 18. (a) Schematics of the wavefield computation problem for anisotropic solid half space, (b) ultrasonic field in water generated by a 1 MHz transducer, (c) pressure and normal stress (σ_{33}) distribution at the fluid–solid interface, (d) displacement distribution in the transversely isotropic solid at the interface and at a distance of 5 mm from the interface. i.e., 10 mm from the transducer face, (e) pressure distribution on a $x - y$ plane at the transducer interface, (f) pressure distribution on an $x - y$ plane at 4.5 mm in front of the transducer closer to the sloid interface.

governing elastodynamic matrix. The governing matrix that is constructed from the boundary and interface conditions can be written as

$$
\begin{bmatrix}
\mathbf{M}_{ss} & \mathbf{M}_{SI} & \mathbf{0} \\
\mathbf{Q}_{IS} & \mathbf{Q}_{II} & \mathbf{S33}_{II*} \\
\mathbf{DF3}_{IS} & \mathbf{DF3}_{II} & -\mathbf{u3}_{II*} \\
0 & 0 & \mathbf{S31}_{II*} \\
0 & 0 & \mathbf{S32}_{II*}
\end{bmatrix}
\left\{
\begin{array}{c}
\mathbf{A}_S \\
\mathbf{A}_I \\
\mathbf{A}_{I*}
\end{array}
\right\}
=
\left\{
\begin{array}{c}
V_{S0} \\
0 \\
0
\end{array}
\right\},
\tag{31}
$$

where $\mathbf{M}$ represents the velocity Green's function matrix in the fluid, $\mathbf{Q}$ represented the pressure Green's function matrix in the fluid and

S33, S31 and **S32** represent the stress Green's function matrix in the transversely isotropic material for σ_{33}, σ_{31} and σ_{32}, respectively. **DF3** is the displacement Green's function matrix in the fluid in the x_3 direction and **u3** is the displacement Green's function in the solid in the x_3 direction. SS means the wavefield on the surface S due to the source layer S, similarly II^* means the respective wavefield at the interface I due to the source layer I^*.

After solving the linear algebra problem in Eq. (31), the source strengths, $\mathbf{A}_s$, $\mathbf{A}_I$ and $\mathbf{A}_{I^*}$, considered in the problem were obtained. Next we computed both the pressure and normal stresses at the fluid–solid interface to check if the boundary conditions are satisfied (Fig. 18(c)). After confirming the solution of the system, the wavefield inside the fluid (in front of the transducer) was calculated as shown in Figs. 18(b), 18(e) and 18(f). In Fig. 18(d), the u_1 and u_3 displacements are plotted at the fluid–solid interface and at a 5-mm distance from the interface. Considering the transversely isotropic material of our interest, the wavefield neither in the solid nor in the fluid are expected to be isotropic in nature. Wave energy at the interface of the transversely isotropic material has clearly bifurcated in two opposite directions, leaving the wavefield with its minimum amplitude at the center aligned with the central axis of the transducer, which is completely opposite to the phenomenon observed in isotropic materials presented earlier[38,45] and supports the physics of orthotropy. To elaborate our understanding and to confirm the phenomena further, the ultrasonic wavefield in the fluid on two $x-y$ planes were plotted (i) near the transducer face and (ii) at a 4.5-mm-distance from the transducer face, which is very close to the interface, respectively. The isotropic wavefield generated at the ultrasonic face (i) Fig. 18(e) gradually became anisotropic near the fluid–solid interface (ii) Fig. 18(f). This is due to the very nature of the transversely isotropic media where the higher wave energy is carried along the x_1 direction compared to the x_2 direction and so does the leaky Lamb waves in the fluid. Together Fig. 18 demonstrates the possibility of numerical computation of the ultrasonic wavefield in the anisotropic solids.

With this numerical computation steps, we conclude this chapter with the promise of DPSM being an important method for calculating ultrasonic field in composite plates with multiple layers. In case of multiple layers, the problem formulation will follow similar steps as it was presented in earlier articles[37,40,45,51] on DPSM but using the anisotropic Green's functions presented in this chapter. Thus, it is possible to compute the full ultrasonic wavefield in any solid media with complex geometry and anisotropy to serve the C-NDE community.

Acknowledgement

The author acknowledges the support of NASA Langley Research Center (LaRC) for this research through Contract no. NNL15AA16C.

References

1. Piascik, B., J. Vickers, D. Lowry, S. Scotti, J. Stewart, and A. Calomino, *DRAFT Materials, Structures, Mechanical Systems, and Manufacturing Roadmap, Technology Area 12*, NASA, Editor. November 2010.
2. Handbook, D.o.D., *Nondestructive Evaluation System Reliability Assessment: Ultrasonic Test System*, in *Appendix C*. 2009 April 7, Department of Defense, USA.
3. Giurgiutiu, V., *Structural Health Monitoring with Piezoelectric Wafer Active Sensors*, 2nd edition. Elsevier Academic Press, 2014.
4. Sachse, W. *et al.*, Recent developments in quantitative ultrasonic NDE of composites. *Ultrasonics* **28**(3) (1990) 97–104.
5. Gresil, M. and V. Giurgiutiu, Prediction of attenuated guided waves propagation in carbon fiber composites using Rayleigh damping model. *Journal of Intelligent Material Systems and Structures* **26** (2015) 2151–2169.
6. Auld, B. A., *Acoustic Fields and Waves in Solids*. Wiley: Department of Defense, USA, 1990.
7. Newberry, B. P. and R. B. Thompson, A paraxial theory for the propagation of ultrasonic beams in anisotropic solids. *Journal of the Acoustical Society of America* **85** (1989) 2290.
8. Mal, A. K., C. C. Yin, and Y. Bar-Cohen, Ultrasonic nondestructive evaluation of cracked composite laminates. *Composites Engineering* **1**(2) (1991) 85–101.
9. Wang, C. Y. and J. D. Achenbach, Elastodynamic fundamental solutions for anisotropic solids. *International Journal of Geophysics* **118** (1994) 384–392.
10. Mouchtachi, A. *et al.*, Ultrasonic study of elastic anisotropy of material composite. *Applied Composite Materials* **11**(6) (2004) 341–351.
11. Moser, B., L. Weber, and A. Mortensen, Damage accumulation during cyclic loading of a continuous alumina fibre reinforced aluminium composite. *Scripta Materialia* **53**(10) (2005) 1111–1115.
12. Li, B. *et al.*, Damage localization in composite laminates based on a quantitative expression of anisotropic wavefront. *Smart Materials and Structures* **22**(6) (2013) 065005.
13. Marguères, P. and F. Meraghni, Damage induced anisotropy and stiffness reduction evaluation in composite materials using ultrasonic wave

transmission. *Composites Part A: Applied Science and Manufacturing* **45** (2013) 134–144.

14. Every, A. G. and K. Y. Kim, Determination of elastic constants of anisotropic solids from elastodynamic Green's functions. *Ultrasonics* **34**(2–5) (1996) 471–472.

15. Hood, J. A. and R. B. Mignogna, Determination of elastic moduli in anisotropic media from ultrasonic contact measurements. *Ultrasonics* **33**(1) (1995) 45–54.

16. Pradhan, B., N. Venu Kumar, and N. S. Rao, Stiffness degradation resulting from 90° ply cracking in angle-ply composite laminates. *Composites Science and Technology* **59**(10) (1999) 1543–1552.

17. Sahay, S. K., R. A. Kline, and R. Mignogna, Phase and group velocity considerations for dynamic modulus measurement in anisotropic media. *Ultrasonics* **30**(6) (1992) 373–382.

18. Sharma, M. D., Group velocity along general direction in a general anisotropic medium. *International Journal of Solids and Structures* **39**(12) (2002) 3277–3288.

19. Siva Shashidhara Reddy, S. *et al.*, Ultrasonic goniometry immersion techniques for the measurement of elastic moduli. *Composite Structures* **67**(1) (2005) 3–17.

20. Vishnuvardhan, J., C. V. Krishnamurthy, and K. Balasubramaniam, Genetic algorithm reconstruction of orthotropic composite plate elastic constants from a single non-symmetric plane ultrasonic velocity data. *Composites Part B: Engineering* **38**(3) (2007) 216–227.

21. Biondi, B., Solving the frequency-dependent Eikonal equation. *62nd Annual International, Soc. Expl. Geophys.* Expanded Abstracts (1992) p. 1351.

22. Pamel, V. A., Sha, G., Rokhlin, S. I., Lowe, M. J. S., Finite-element modelling of elastic wave propagation and scattering within heterogenous media. *Proceedings of Royal Society A*, (2017).

23. Shaw, R. P., Boundary integral equation methods applied to wave problems, In Banerjee, P. K. and R. Butterfield, (Ed.), *Developments in Boundary Element Methods — 1*, Applied Science Publishers, Department of Defense, USA, 1979, pp. 121–153.

24. Zhao. X. and J. L. Rose, Boundary element modeling for defect characterization potential in a wave guide. *International Journal of Solids and Structures* **40**(11) (2003) 2645–2658.

25. Sánchez-Sesma, F. J. and M. Campillo, Diffraction of P, SV and Rayleigh waves by topographic features: A boundary integral formulation. *Bulletin of the Seismological Society of America*, **81** (1991) 2234–2253.

26. Sánchez-Sesma, F. J. and M. Campillo, Topographic effects for incident P, SV and Rayleigh waves. *Tectonophysics*, **218** (1993) 113–125.

27. Pointer, T., E. Liu and J. A. Hudson, Numerical modeling of seismic waves scattered by hydrofractures: Application of the indirect boundary element method. *International Journal of Geophysics* **135** (1998) 289–303.

28. Bouchon, M. and F. J. Sánchez-Sesma, Boundary integral equations and boundary elements methods in elastodynamics, In Wu, R. S., X. B. Xie

and X. Y. Wu (Ed.), *Advances in Wave Propagation in Heterogeneous Earth, Advances in Geophysics*, Elsevier-Academic Press, New York, Boston, 2007.

29. Wen, J. J. and M. A. Breazeale, A diffraction beam field expressed as the superposition of Gaussian beams. *Journal of the Acoustical Society of America* **83** (1988) 1752.

30. Spies, M., Transducer field modeling in anisotropic media by superposition of Gaussian base functions. *Journal of the Acoustical Society of America* **105** (1999) 633.

31. C. Rajamohan, and J. Raamachandran, Bending of anisotropic plates by charge simulation method. *Advances in Engineering Software* **30**(5) (1999) 369.

32. Ballisti, C. H., The multiple multipole method (MMP) in electro and magnetostatic problems. *IEEE Transactions on Magnetics* **19**(3) (1983) 2367.

33. Hafner, C., MMP calculations of guided waves. *IEEE Transactions on Magnetics* **21** (1985) 2310.

34. Imhof, M. G., Computing the elastic scattering from inclusions using the multiple multipoles method in three dimensions. *International Journal of Geophysics* **156** (2004) 287.

35. Moll, J., C. Rezk-Salama, R. T. Schulte, T. Klinkert, C.-P. Fritzen, and A. Kolb, Modelling of wave-based SHM systems using the spectral element method, under interactive simulation and visualization of lamb wave propagation in isotropic and anisotropic structures. *Journal of Physics: Conference Series* **305** (2011) 012095.

36. Leckey, C. A., M. D. Rogge, and F. R. Parker, Guided waves in anisotropic and quasi-isotropic aerospace composites: Three-dimensional simulation and experiment. *Ultrasonics* **54** (2014) 385–394.

37. Banerjee, S. and T. Kundu, Ultrasonic field modeling in plates immersed in fluids. *International Journal of Solids and Structures* **44**(18–19) (2007) 6013–6029.

38. Banerjee, S., N. Alnuaimi, and T. Kundu, DPSM technique for ultrasonic field modelling near fluid-solid Interface. *Ultrasonics* **46**(3) (2007) 235–250.

39. Banerjee, S. and T. Kundu, Semi-analytical modeling of ultrasonic fields in solids with internal anomalies immersed in a fluid. *Wave Motion* **45**(5) (2008) 581–595.

40. Banerjee, S. and T. Kundu, Elastic wave field computation in multilayered non-planar solid structures: A mesh-free semi-analytical approach. *Journal of Acoustical Society of America* **123**(3) (2008) 1371–1382.

41. Banerjee, S., S. Das, T. Kundu, and D. Placko, Controlled space radiation concept for mesh-free semi-analytical technique to model wave fields in complex geometries. *Ultrasonics* **48**(8) (2009) 615–622.

42. Banerjee, S. and T. Kundu, Elastic wave propagation in sinusoidally corrugated waveguides. *Journal of the Acoustical Society of America* **119**(4) (2006) 2006–2017.

43. Rahani, E. K. and T. Kundu, Gaussian-DPSM (G-DPSM) and element source method (ESM) modifications to DPSM for ultrasonic field modeling. *Ultrasonics* **51**(5) (2011) 625–631.

44. Rokhlin, S. I., D. E. Chimenti and P. B. Nagy, *Physical Ultrasonics of Composites*. Oxford University Press, Oxford, UK, 2011.

45. Banerjee, S. and T. Kundu, *Advanced Application of Distributed Point Source Method — Ultrasonic Field Modeling in Solid media*, In T. Kundu and D. Placko (Eds.), *DPSM for Modeling Engineering Problems*, John & Wiley Publication, Hoboken, New Jersey, USA, 2007.

46. Every, A. G. *et al.*, Phonon focusing caustics in crystals and their diffraction broadening at ultrasonic frequencies. *Transactions of Royal Society of South Africa* **58**(2) (2004) 119–128.

47. Kim, K. Y., A. G. Every, and W. Sachse, Focusing of fast transverse modes in (001) silicon at ultrasonic frequencies. *Journal of Acoustical Society of America* **95**(4) (1994) 1942.

48. Pluta, M., *et al.*, Angular spectrum approach for the computation of group and phase velocity surfaces of acoustic waves in anisotropic materials. *Ultrasonics* **38** (2000) 232–236.

49. Tverdokhlebov, A. and J. Rose, On Green's function for elastic waves in anisotropic media. *Journal of Acoustical Society of America* **83** (1988) 844–870.

50. Yeatts, F. R., Elastic radiation from a point force in an anisotropic medium. *Physics Review B* **29** (1984) 1674–1684.

51. Banerjee, S. and T. Kundu, Symmetric and anti-symmetric Rayleigh-Lamb modes in sinusoidally corrugated waveguides: An analytical approach. *International Journal of Solids and Structures* **43** (2006) 6551–6567.

Chapter 4

Degradation Detection in Composite Structures with Piezoelectric Transducers

Wiesław M. Ostachowicz*†, Paweł H. Malinowski†,
and Tomasz Wandowski†

*Warsaw University of Technology,
Faculty of Automotive and Construction Machinery,
Warsaw, Poland
† Institute of Fluid-Flow Machinery, Polish Academy of Sciences
14 Fiszera Street, 80-231, Gdansk, Poland
*wieslaw@imp.gda.pl

This chapter presents results of the assessment of composites using techniques based on piezoelectric (PZT) transducers. These transducers were used as the active elements in the electromechanical impedance method and guided wave-based propagation method. The second method was used together with scanning vibrometer, which allowed to visualize the propagation of elastic waves. Small samples as well as real structural elements were investigated. Effective methods for adhesive bonds and damage assessment were investigated. These methods were based on special signal processing techniques that allowed to identify signal anomalies caused by degradation of the structure. The obtained results showed that both methods are useful for the assessment of adhesive bonding and damage detection.

1. Introduction

Fiber-reinforced composite materials are nowadays increasingly used in many branches of industry. They are made in the form of polymer matrix with embedded reinforcing fibers. Most popular are polymers reinforced by carbon, glass or aramid fibers [carbon fiber reinforced polymer (CFRP), glass fiber reinforced polymer (GFRP), aramid fiber reinforced polymer (AFRP)]. These materials are also used as a part of structural composites. Structural composites consist of skin made from fiber-reinforced polymer with, for example, honeycomb core. They are widely utilized in aerospace, automotive or maritime industries. The main advantage of these materials is their high mechanical strength to density ratio. These materials can be easily formed and are very resistant to corrosion or chemical degradation. But the main disadvantage is their high vulnerability to initiation of hidden delamination caused by impact load. Delamination is initiated due to inter-laminar shear stresses existing between layers of composite under the load acting transversely to the surface. Moreover, they are generally resistant to chemical degradation (especially composites based on epoxy resin); however, due to application of high temperature, chemical reactions can occur that can lead to degradation of the mechanical performance of these materials. Source of such high temperatures can be, for example, exhausting gases from turbojet engines or lightning strike. Besides chemical degradation, high temperature can also lead to initiation of delamination. Based on these properties, the process of composite material degradation can be divided to mechanical degradation and chemical degradation. Mechanical degradation is caused by mechanical load (in the impact or static form). It can be also caused by humidity and water ingress in the composite material. Water can be frozen and as a consequence it can lead to a local increase of stresses ending with delamination. Chemical degradation can be caused by the above-mentioned exposition of material to high temperatures (higher than temperature of glass transition). Chemical degradation can also be caused by aggressive chemical substances like acids. In the composite aerospace structures, such an aggressive fluid can be Skydrol (fluid utilized in airplane hydraulic power installation). When Skydrol mixes with water, two chemical phases are created, one of which is very aggressive. Moreover, deicing fluids utilized in aerospace are also aggressive.[1] It should be mentioned that a very long exposition of composite material to water can change the mechanical performance of glass-reinforced composites.[1] If the water contains salt, the probability of degradation of the composite material is high. Composite materials are also vulnerable to ultraviolet radiation, which can also be a source of chemical degradation.[2]

Composite structural elements may be joined together by adhesive bonding. The joints between composite elements should ensure safe exploitation of a structure. The reliability of bonded joints is not high still. Due to requirements of certification processes for composite structures, the crucial parts (e.g., segments of fuselage) still need to be riveted together. Bonding agents can be utilized during the manufacturing process as well as during the process of structural repair (e.g., adhesively bonded patch). The performance and reliability of adhesive bonds joining structural elements depends strictly on the physical–chemical properties of the adhered surfaces. The physical–chemical properties depend on the process of surface preparation of the composite elements. An improper process of surface preparation may lead to a reduction in the performance of the bonded joint. Such joints with reduced performance are called weak bonds. Exploitation of the composite structure with weak bonds may simply lead to its failure with terrible results. The research of the assessment of the performance of bonded joints can be done in two ways.[3] The first one is related to an assessment of quality of the surface prepared before the bonding process. The second way is an assessment of the performance of the bond itself.

It should be emphasized here that weak bonds can be source of improper manufacturing process as well as result of normal in-service exploitation of the composite structure. For example, contamination of the release agent may come from manufacturing through contamination of the fuel. Skydrol hydraulic fluid or deicing fluid can have an influence on the bonded joints during the exploitation process.[4] Moisture contamination, which may result from the manufacturing process (during ultrasonic non-destructive testing (NDT) with water-coupled transducers) or due to exposure to water during the in-service operations, also has an important influence. The moisture located in the composite may freeze (e.g., at cruising flight level) and may turn a potential source of delamination.

The results of destructive mechanical testing show that the reduced performance of the bonded joint can be caused by the above-mentioned contamination as well as by improperly conducted process of adhesive curing. Markatos *et al.*[5] presented results of research related to influence of pre-bond release agent and moisture contamination on mode I fracture toughness of bonded joints of carbon-reinforced composites. The results show that contamination is the source of a 60% drop in the I-mode energy release rate, called G_{IC}, in relation to reference bond for uncontaminated samples. Markatos *et al.*[6] also investigated the influence of improperly conducted adhesive curing process. A reduction of the recommended curing temperature leads to nearly 95% drop in G_{IC} in relation to reference samples.

The physical–chemical degradation of composites presented here show that structures manufactured using these materials need to be assessed using NDT or structural health monitoring (SHM) techniques. Application of these techniques allow to improve the safety of in-service exploitation of composite structures. In the following sections, select NDT and SHM techniques based on piezoelectric (PZT) transducers will be presented.

2. Adhesive Bond Assessment using EMI Method

The electromechanical impedance (EMI) method is considered as one of the promising solutions that can be used for NDT or structural health monitoring (SHM). In the EMI method, a PZT sensor bonded to the inspected structure is used (Fig. 1). The sensor excites and senses the structure's response due to PZT effect. For an assessment of the structure's condition, resistance, reactance, impedance and conductance are measured. These quantities are expressed as a function of frequency and measured by an impedance analyzer (Fig. 1). If there is a damage, it may cause a change in the number of resonance peaks in a given bandwidth, peak shift, or magnitude change. These changes can be treated as an indicator of defect of the host structure.[7] Moreover, degradation of sensor bonding has an influence on the EMI method response.[8–10] Generally, the imaginary part of the electromechanical impedance (or admittance) measurement can be used for sensor self-diagnostics while the real part can be utilized for damage detection purposes (e.g., Ref. [11]).

Damage-related features may be present in various EMI spectra, depending on the considered case. These bands depend on the inspected structure and the used PZT sensor. A comparison of damage sensitivity of high- and low-frequency bandwidths was presented in Ref. [12]. Moreover Park et al.[12] underline the good sensitivity of the EMI method based on analysis of the bandwidth near frequency of thickness vibration mode of the PZT sensors. This frequency lies in higher frequency bands.

The PZT sensor used in this research is manufactured using SONOX P502 material with a thickness resonance around 4 MHz. This resonance is visible in the admittance ($|Y|$) and conductance (G) bands as a strong peak. The susceptance (B) has a zero-crossing at this point. The relation

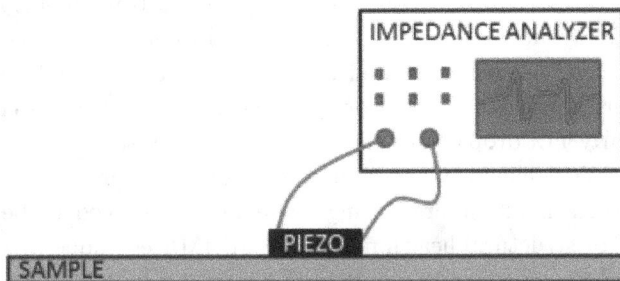

Figure 1.   Schematic drawing of an EMI method with inspected sample, PZT sensor for excitation and sensing, and an impedance analyzer for conducting measurements.

between these three quantities as follows:

$$Y = G + iB, \tag{1}$$

$$|Y| = \sqrt{G^2 + B^2} \tag{2}$$

The goal of this research is to investigate if any correlation exists between adhesive bond condition and the EMI spectra. Carbon fiber-reinforced plastic/polymer (CFRP) samples undergo pre-bond modification. The samples are comprised of a 1.5-mm thick and 2.6-mm thick plates joined with a film adhesive (Cytec, FM300K.05). The planar dimensions of the samples are 100 mm × 50 mm (MO) and 100 mm × 100 mm (CS). The sensors are bonded in the middle of each sample face.

Two cases were considered. In the first case, one of the adherend plates was contaminated with moisture and then bonded to an uncontaminated plate. Four moisture cases were investigated, each with different level of moisture obtained by keeping the plates in environmental chamber with controlled humidity. The level of contamination was expressed by mass increase related to amount of moisture intake (Table 1). The ascending numbering of the samples MO1–MO4 corresponds to the ascending contamination level. In the second case, one of the plates was bonded with lowered temperature of adhesive curing. The proper curing temperature is in the interval 170–180°C. The considered lowered values were 150°C (symbol: CS150) and 160°C (symbol: CS160).

In order to quantitatively assess the adhesive bonds, a numerical index was proposed. This index is based on calculation of the Frechet distance (F) between the two curves. The curves represent the admittance characteristics on a complex plane for the 3–5 MHz band. The Frechet distance is the minimum distance required to connect two points constrained on two separate paths, as the points travel without backtracking along their

Table 1.   Considered adhesive bond cases with moisture contamination.

Symbol	Prebond mass increase due to moisture (%)
MO1	0.45
MO2	0.80
MO3	1.13
MO4	1.25

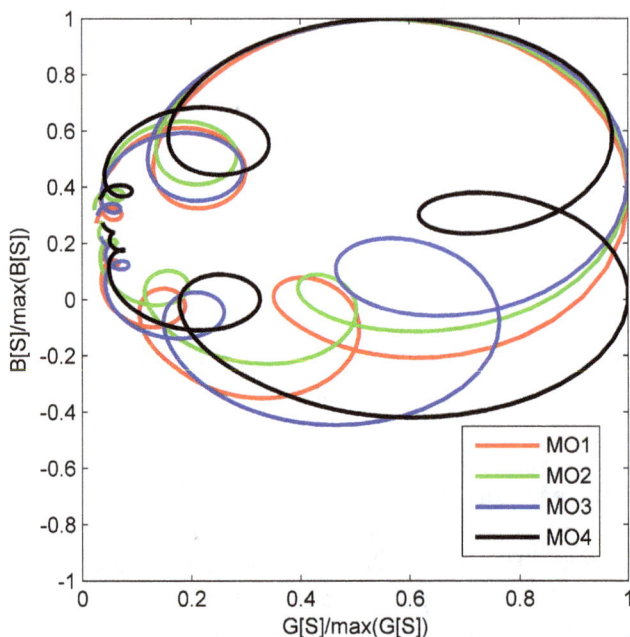

Figure 2. Admittance characteristics for the bonded samples with moisture contamination.

respective curves from one endpoint to the other.[13,14] The definition is symmetric with respect to the two curves.

The admittance characteristic of the MO samples are presented in Fig. 2 on a complex plane. The horizontal axis (real) represents conductance (G), while the vertical axis (imaginary) represents the susceptance (B). Both quantities were normalized to their respective maximum values in order to facilitate the comparison of the shape of the curves. In order to quantitatively assess the adhesive bond, the Frechet distance was calculated between the curves. The promising results of using this distance for the assessment of adhesive bonds contaminated with release agent were presented in Ref. [15]. The Frechet distance (F) was calculated for the curves MO1–MO4 presented in Fig. 3. The F values inform about the dissimilarity of the curves. The results were plotted in color scale in Fig. 3.

The distance between the same curves is zero. One can notice that there are differences among the tested cases. Let's take the case with the lowest contamination (MO1) as a reference. It can be observed that the F value for MO2–MO4 increases in correlation with the increase in the contamination

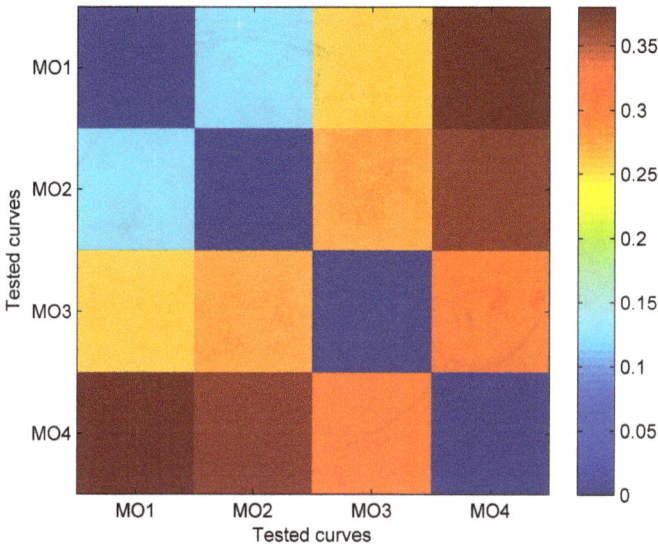

Figure 3. Results (Frechet distance) of comparison of admittance characteristics for the bonded samples with moisture contamination.

level. The results indicate that the admittance curves become more and more different as the bondline contamination level increases.

In the case of the lowered temperature of adhesive curing, the samples were measured together with a reference (REF) sample. The REF sample had adhesive bond cured at the proper temperature. The obtained admittance curves are depicted in Fig. 4. The shape differences are clearly visible and in the next step they were quantified by calculation of the Frechet distance between the considered set of curves. The results are depicted in Fig. 5. The F value increases with decreasing adhesive curing temperature. The sample with bond cured at $160°\,$C has higher F value than the REF and the sample cured at $150°\,$C differs even more from the REF.

3. Damage Detection using EMI Method

The EMI method described in Section 2 and applied for the assessment of adhesive bonds is also under development for the purposes of damage detection. The methodology of using Frechet distance was continued for the damage detection purposes. Again the 3–5 MHz frequency band was considered for the calculation and comparison of the admittance curves.

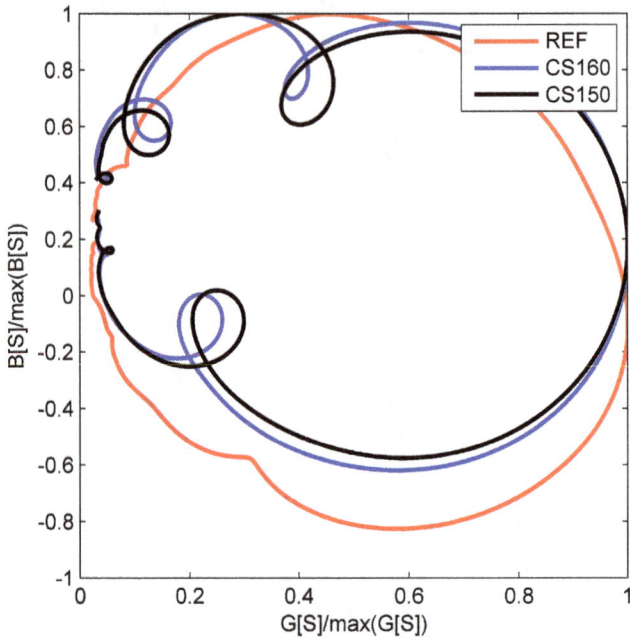

Figure 4.   Admittance characteristics for the bonded samples with lowered adhesive curing temperature.

The investigated object was a 500 mm × 500 mm plate made of glass fiber-reinforced polymer (GFRP sample). The stacking sequence of the layers was $[0/90/0/90]_s$. The sample was 3 mm thick. In order to simulate damage, a square (10 mm × 10 mm) Teflon tape was inserted between the second and the third layer. The teflon was located in the right lower corner 120 mm away from the two closest edges (Fig. 6). A PZT sensor was attached at the sample surface at the same x, y position. This sensor was denoted as "G teflon" (see Fig. 7).

Three additional sensors were attached for reference purposes. The three sensors were located symmetrically at the remaining three corners of the plate (Fig. 6). These sensors were denoted as G1–G3 and located at undamaged areas of the plate.

The results of calculating of the Frechet distance are presented in Fig. 8 The sensor located at the artificially made damage was clearly distinguished from the remaining cases. The differences between the three reference cases

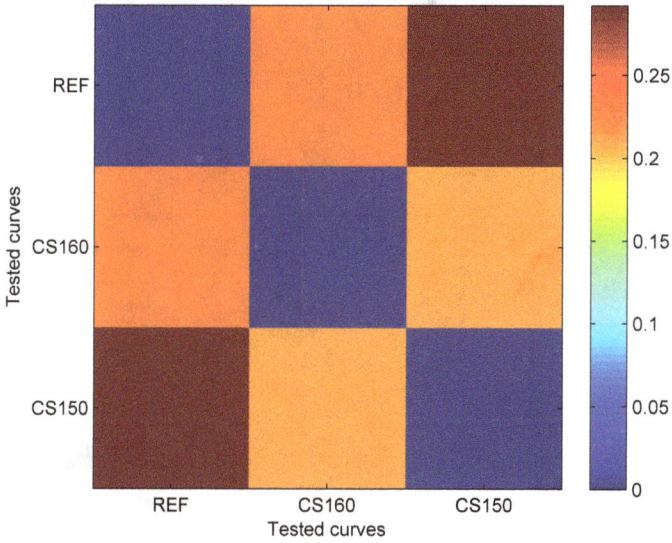

Figure 5. Results (Frechet distance) of the comparison of admittance characteristics for the bonded samples with lowered adhesive curing temperature.

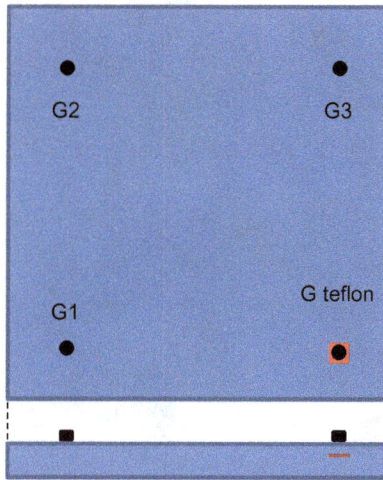

Figure 6. A Schematic drawing of the investigated GFRP sample with four PZT sensors and a teflon insert collocated with one of the sensors.

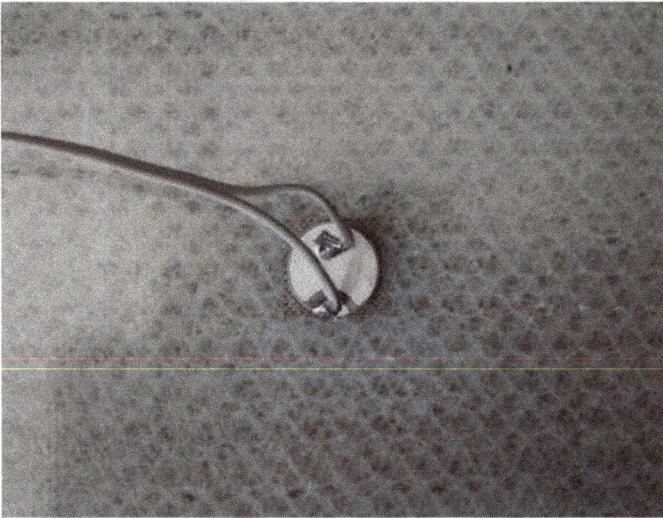

Figure 7. A photo of PZT sensor (G teflon) bonded at the location of teflon insert.

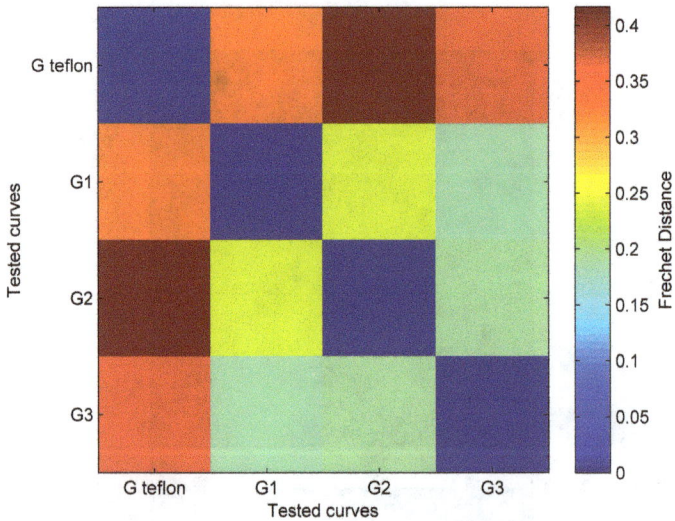

Figure 8. Results (Frechet distance) of comparison of admittance characteristics for the detection of teflon insert at the location of sensor denoted as "G teflon".

(G1–G3) are lower than those between 'G teflon' results and the sensors G1–G3. Teflon detection was successful.

4. Damage Detection using Elastic Wave Propagation

Elastic waves can propagate in infinite solids as longitudinal and shear waves. In a finite solid, for example in a plate, which is limited by two parallel surfaces, guided wave propagation is observed. Guided waves are created as a result of internal reflections of longitudinal and shear waves. The guided wave based method utilizes the fact that any form of discontinuities in elastic media is a potential source of change of wave propagation. These changes in wave propagation can be observed as wave reflections, scattering and mode conversions. The initiated damage leads to stiffness change and as a consequence causes changes in wave propagation. Observations of changes in guided wave propagation allow to detect and localize the damage. This method can be exploited as passive or active method. In the passive approach, signals are only registered by sensors located in the structure. In the active method, actuators are utilized for guided wave excitation whereas sensors register the signals.[16] In both approaches, PZT transducers are commonly utilized. Due to the direct and the converse PZT effects, these transducers can sense and excite guided waves. PZT transducers are located on the interrogated structure and are arranged in networks with different topologies. They can be grouped in concentrated or distributed networks. In the concentrated networks, sensors are arranged in linear, circular, cross-like, or square-like array.[17–20] This approach is called point-wise method because signals are taken from the array of small numbers of single measurement points. Many signal processing techniques have been proposed in order to extract damage localization. A very popular technique is the delay-and-sum algorithm with different modifications and improvements (e.g., Refs. [21, 22]). More information about signal processing for damage detection and localization in composite materials can be found in Ref. [23].

A very efficient approach for wave sensing is based on scanning laser Doppler vibrometry (SLDV). This is a non-contact approach where the waves are excited by a PZT transducer and laser vibrometer is utilized for its sensing. A great advantage is the possibility of non-contact measurements in chosen points of the structure. It needs to be emphasized here that measurements can be taken from very dense mesh of measuring points covering the whole surface of the interrogated structure. This allows to create the animation of guided wave propagation in the structure. This approach is called full wave-field method and is especially suitable for wavelength filtering and wavelength analysis and observation of guided wave mode conversion phenomenon.[24–26] In this section, a few examples of damage detection results using laser vibrometry will be presented.

The first case is related to small GFRP sample with delamination. The sample dimensions were 100 mm × 100 mm × 3.3 mm. It was manufactured using 12 plies of glass S-fibers with a stacking sequence of $[0/90/0/90/0/90]_s$ and Araldite LY1564 epoxy. Guided waves were excited using PZT transducer in the form of a disk with 10 mm diameter and 0.5 mm thickness. Transducer was made of NOLIAC NCE51 piezoelectric material. This type of PZT transducer was used in all examples presented in this section. Guided wave sensing was performed by laser vibrometer for a dense mesh of measuring points covering the whole surface of the sample. Next, for all gathered signals, RMS index was calculated. This index is defined as follows:

$$\text{RMS}_i = \sqrt{\frac{1}{N} \sum_{j=1}^{N} S_{i,j}^2}, \tag{3}$$

where $S_{i,j}$ is the signal gathered in the point i and N the length of the signal. Using the RMS index values, a map of energy of propagating waves can be created. In Fig. 9, RMS energy map for excitation frequency of 100 kHz was presented. Elastic wave energy concentrates around the PZT transducer, generating waves. Moreover, energy concentrates in the region of delamination denoted as D. The location of delamination can be clearly distinguished. This energy concentration is related to elastic wave entrapment in the delamination.

In the next case, a larger sample with dimensions 200 mm × 600 mm × 3.5 mm was investigated. This time sample was manufactured using CFRP pre-pregs. This sample was manufactured with eight layers of GG204P IMP503 42 (balanced fiber Plain 3 K fabric) pre-pregs. The CFRP sample surface was prepared before measurements by covering it with a retro-reflective tape. Guided waves were excited using PZT transducer while the sensing was performed by laser vibrometer. In this case, much higher excitation frequency than the previous case was utilized. The excitation

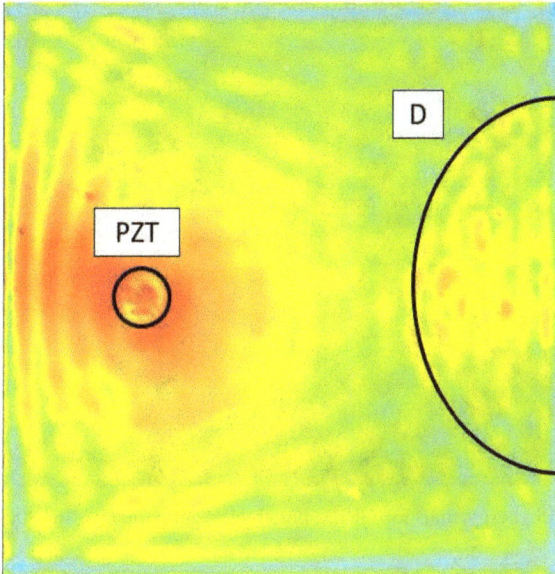

Figure 9.   RMS energy map for small GFRP sample with delamination for excitation frequency 100 kHz.

frequency was equal to 300 kHz. In Fig. 10, results in the form of RMS energy map is presented. It should be mentioned that the result is shown for only one half of the sample surface (left). In this part, two PZT transducers were bonded to the surface (PZT1 and PZT2 in Fig. 10) but only PZT2 was utilized for guided wave excitation. Transducer PZT1 was not utilized in measurements reported here. Strong energy concentration around PZT2 generating elastic waves can be noticed. Wave energy has also concentrated around PZT1, which is only an additional mass here. Moreover, waves are strongly damped in the region of delamination denoted as *D*. The location of delamination can be clearly distinguished. It should be emphasized here that for higher frequencies, different types of wave interaction with delamination is observed. Here a strong damping (energy reduction) is observed whereas for lower frequencies, wave entrapment and energy concentration was observed.

The second example is related to tapered GFRP panel with dimensions: 500 mm × 600 mm. Panel thickness varied from 1.1 to 4.1 mm. The panel was manufactured using different numbers of VV 192T 202 IMP503 layers.

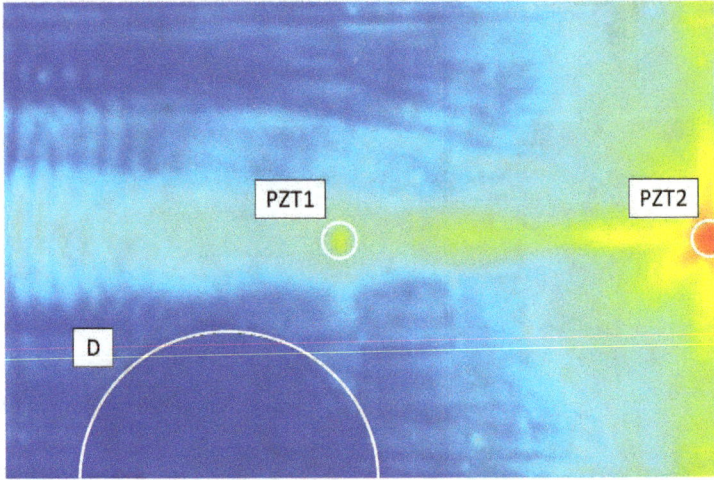

Figure 10.   RMS energy map for CFRP panel with delamination for 300 kHz excitation frequency.

Between internal layers of the composite panel, square teflon insert was placed in order to simulate delamination. The edge of this insert was equal to 20 mm. Teflon insert locations is visible in the results. This method represents a less realistic form of damage, unlike in the previously presented case. However, by placing the teflon insert, it is much easier to control the size, shape and location of the simulated delamination. Moreover, it is very easy to place such delamination between chosen layers (simulation of symmetrical and non-symmetrical placement according to the sample thickness).

Similar to the previous case, GFRP sample surface was prepared before measurements by covering it using retro-reflective tape. Guided waves were excited using a PZT transducer.

In Fig. 11, RMS energy maps for different guided wave excitation carrier frequencies are presented. In each case, tone-burst Hannnig window modulated signal with five sine cycles was utilized.

In the presented energy maps (in Fig. 11), strong energy concentration in the middle of the panel is visible. This energy concentration is related to guided wave excitation by the PZT transducer. This energy concentration in a small square area located in the right upper quarter of panel can be observed in all the cases. This is related to guided wave interaction with teflon insert located between internal layers of the composite panel. The location of the simulated delamination is clearly visible in all the

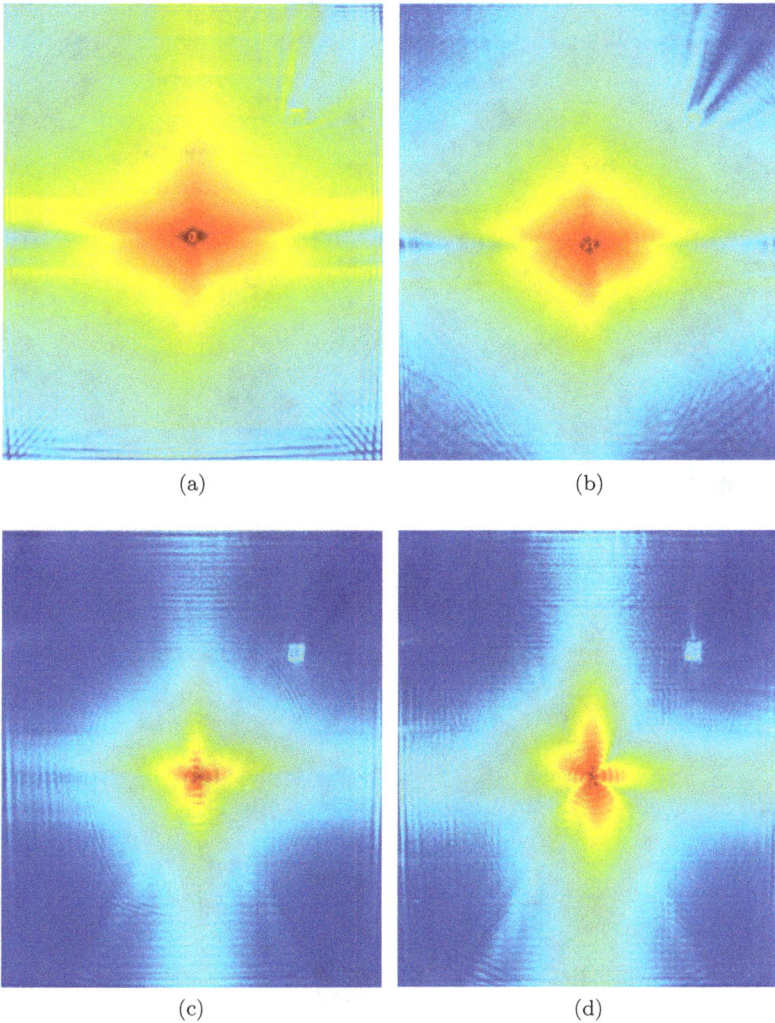

Figure 11. RMS energy maps for GFRP tapered panel for excitation frequency:
(a) 50 kHz, (b) 100 kHz, (c) 150 kHz and (d) 200 kHz.

cases. Comparing results of the energy distribution for all excitation
frequencies, it can be observed that for higher frequencies waves are damped
stronger.

In Fig. 12, the chosen frames from animations of guided wave
propagation for different excitation frequencies in GFRP panel with
simulated delaminations are presented. In all of the presented results,

(a)   (b)

(c)

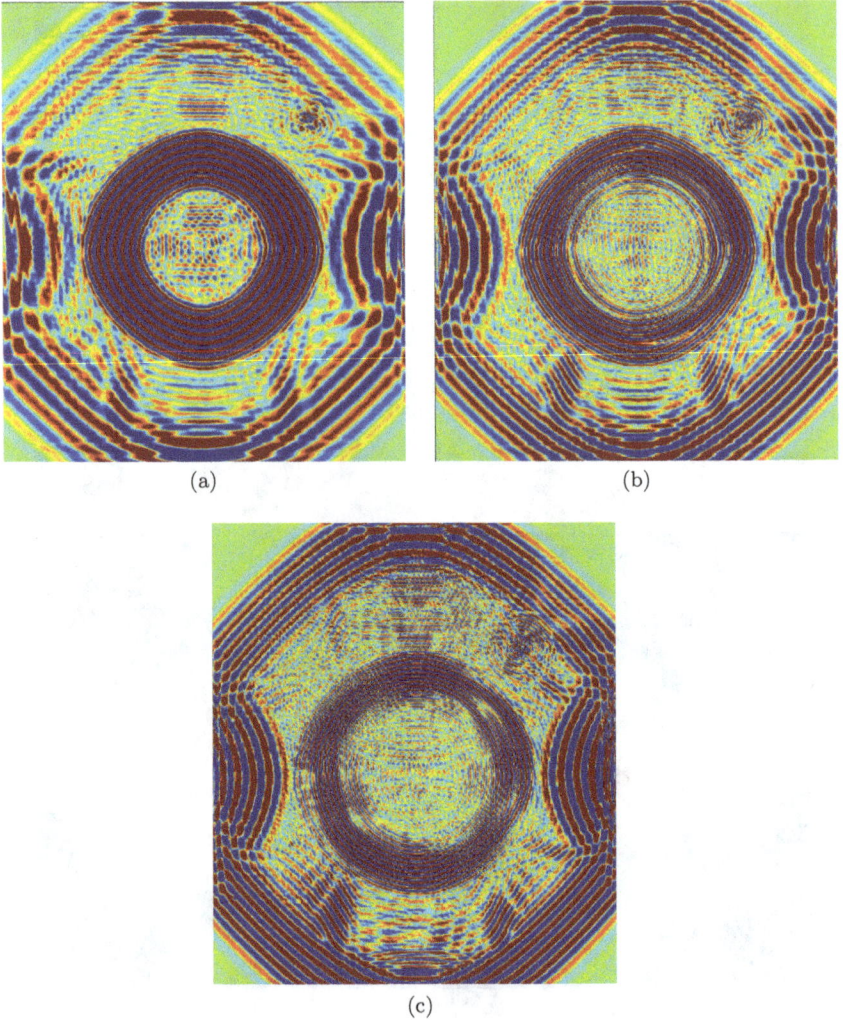

Figure 12.   Guided wave propagation in GFRP panel with simulated delamination for excitation frequencies: (a) 100 kHz, (b) 150 kHz and (c) 200 kHz.

propagation of two fundamental guided wave modes can be noticed. Wavelength and velocity of propagation of both modes differ. The wavelength of faster mode (S_0) is larger while it is smaller for the slower propagating mode (antisymmetric A_0). As it was already mentioned, thickness of the panel varies. In Fig. 12, this thickness varies in vertical direction from thicker section (bottom panel edge) to thinner section (top panel edge). Variation of panel thickness results in variation of the propagation velocities of both the guided wave modes. This effect can be observed for S_0 mode especially in the case of excitation frequency 200 kHz (Fig. 12(c)). This is clearly visible as the S_0 mode achieves the bottom panel edge faster than the top edge. The results of detailed analysis of guided wave propagation in tapered composite plate can be found in Refs. [27, 28].

In the results presented in Fig. 12, another interesting effect is visible — mode conversion. Faster mode S_0 due to interaction with simulated delamination converts to A_0 mode. After conversion, propagation of the new mode with wavelength comparable to the excited A_0 mode is visible. This effect is observed for all of the results presented in Fig. 12 and can be utilized for the purposes of damage detection and localization.

In the next step, the results of the guided wave propagation in GFRP panel with cut-out hole for a window frame (part of helicopter) is presented. This panel is manufactured using four layers of ECST-55 glass fiber reinforcement. The area around the window opening is curved. Moreover, two circular teflon inserts with diameters 20 and 30 mm were located symmetrically between the two layers (with respect to the panel thickness). GFRP sample surface was prepared before measurements. It was covered by retro-reflective tape. The results of the guided wave propagation of this panel for excitation frequencies 15 and 50 kHz are presented. Analyzing both results, propagation of A_0 and S_0 modes can be observed. Moreover, in Fig. 13, the conversion of S_0 mode to A_0 due to interactions with both sizes of simulated delamination (denoted as D) can be observed. In addition, mode conversion S_0 to A_0 can be observed in the region near the window opening where the stiffener is formed (top edge in Fig. 13). It should be mentioned that only the area near the bottom edge of the window opening is measured.

In this case, results for lower excitation frequencies than in previous case are presented in order to show the mode conversion effects.

It should be emphasized here that decision on excitation frequency is very important from the point of view of damage detection and localization process. High excitation frequency results in high precision in damage location using which even a very small damage can be detected (small wavelength). However, damping of guided waves increases with frequency and distance of the guided wave propagation decreases. This is a compromise between the achievable distance of guided wave propagation

(a)

(b)

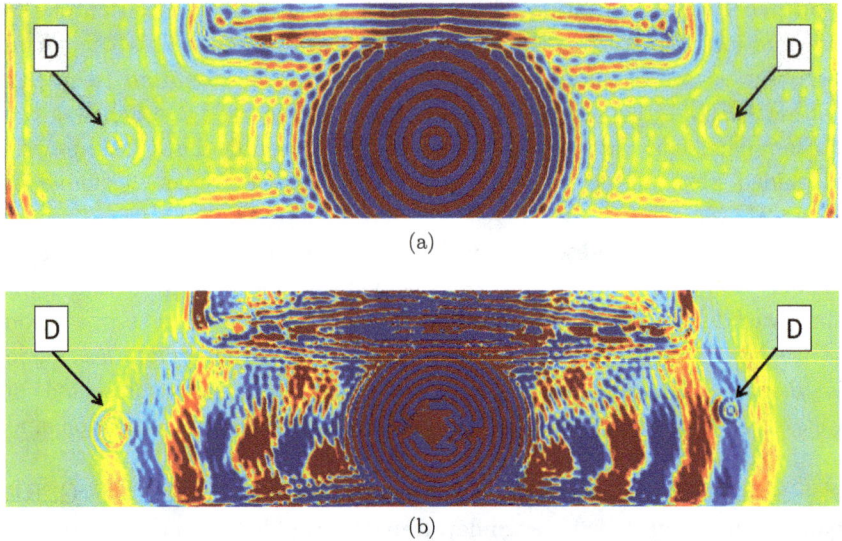

Figure 13.   Guided wave propagation in GFRP panel with two simulated delaminations for excitation frequency of (a) 15 kHz and (b) 50 kHz.

(sensing area where damage can be detected) and its resolution (smallest dimension of damage that can be detected).

The last case investigated in this section is related to CFRP panel with omega stiffeners. This panel consists of flat CFRP panel bonded to a panel formed with omega stiffeners. The number of INTERGLASS 98141-Twill2/2 carbon layers in each part was three (six in total). In this panel, delamination was simulated by teflon insert with a diameter 30 mm, placed between internal layers of the flat panel. Two excitation frequencies were used: 50 and 100 kHz. Measurements were taken on both sides of the panel: first measurement from the stiffeners side and without any surface preparation, and second measurements from the flat side covered by retro-reflective tape. The results of the guided wave propagation for the side with stiffeners for frequencies 50 and 100 kHz are presented in Figs. 14 and 15, respectively. In both cases, two selected frames from animation of guided waves propagation are presented. The whole panel area was not measured by laser vibrometer due to the stiffeners in the panel (see Figs. 14 and 15). Analyzing the presented results, propagation of two guided wave modes A_0 and S_0 can be observed in Figs. 14(a) and 15(a). Moreover, due to the interaction of symmetric mode S_0 with stiffeners, this mode converts to A_0 mode, which can be observed in Figs. 14(b) and 15(b). Due to mode

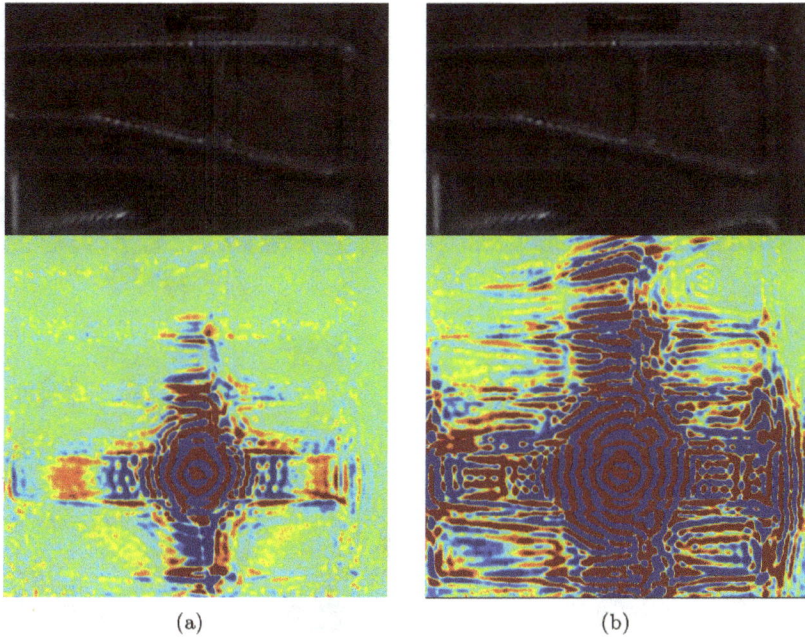

Figure 14. Guided wave propagation in CFRP stiffened panel with simulated delamination for excitation frequency of 50 kHz (measurements taken at stiffeners side).

conversion effects in guided wave propagation, the location of the stiffeners can be noticed.

Based on the gathered measurements, the RMS energy maps for both excitation frequencies are created and presented in Fig. 16. Analyzing these results, strong energy concentration in the location where guided wave excitation was applied can be observed. On analyzing energy distribution, the location of the stiffeners and simulated delamination (denoted as D) can be distinguished. The locations of the simulated delaminations in both cases are clearly visible.

The wave propagation around the delamination in Figs. 14(b) and 15(b) show the mode conversion effect. This is conversion of mode S_0 to A_0.

Similar measurements of guided wave propagation were taken on the flat side covered by retro-reflective tape. The results of these measurements for excitation frequencies 50 and 100 kHz are presented in Figs. 17 and 18. Three time frames from animation of guided waves are selected for each excitation frequency. In the first frames (see Figs. 17(a) and 18(a)), propagation of A_0 and S_0 modes can be observed. Due to the interaction of

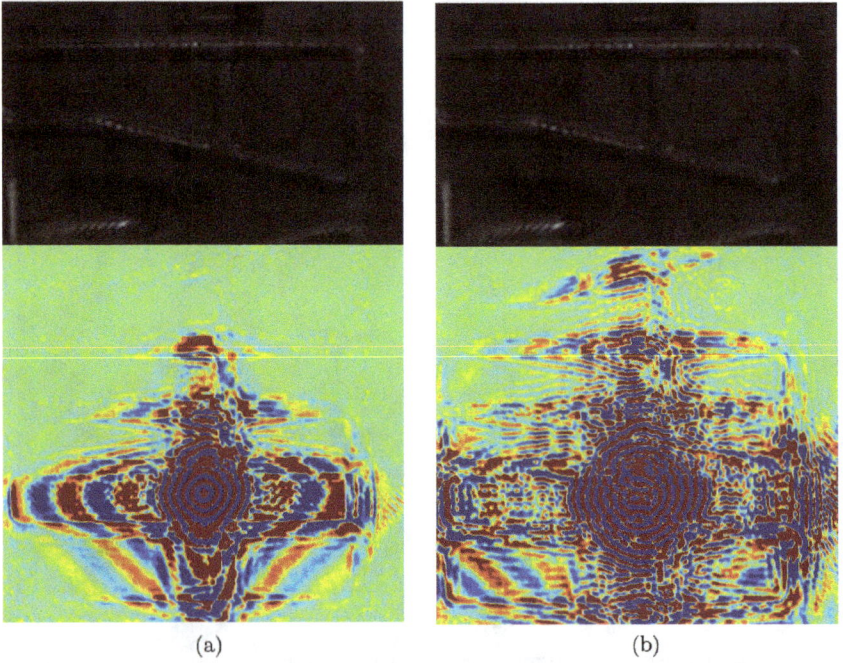

Figure 15. Guided wave propagation in CFRP stiffened panel with simulated delamination for excitation frequency of 100 kHz (measurements taken at stiffeners side).

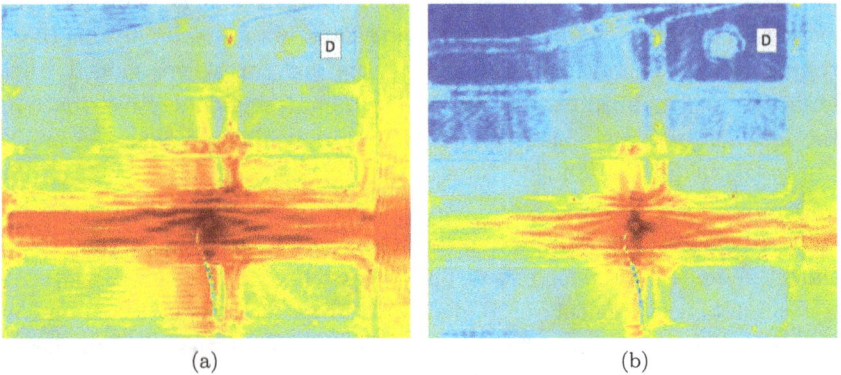

Figure 16. RMS energy maps for CFRP stiffened panel with simulated delamination for excitation frequency: (a) 50 kHz and (b) 100 kHz.

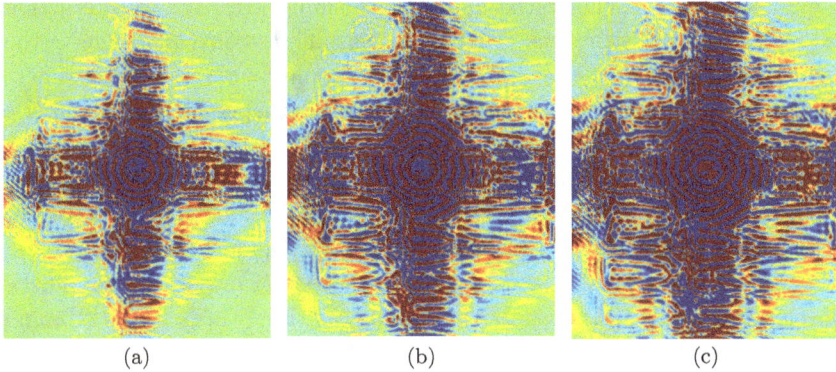

Figure 17. Guided wave propagation in CFRP stiffened panel with simulated delamination for excitation frequency 50 kHz (measurements taken at flat side).

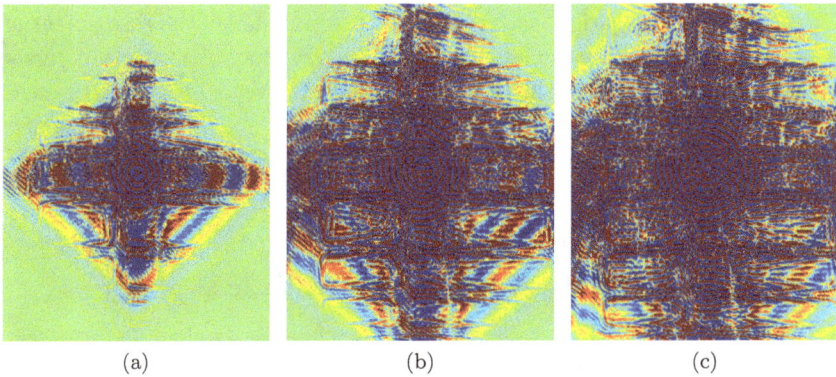

Figure 18. Guided wave propagation in CFRP stiffened panel with simulated delamination for excitation frequency 100 kHz (measurements taken at flat side).

S_0 mode with stiffeners, mode conversion to A_0 mode occurs. This is visible in all the frames presented in Figs. 17 and 18. Moreover, the conversion of mode S_0 to A_0 due to its interaction with simulated delamination can be observed in the top left corners of Figs. 17(b), 17(c), 18(b) and 18(c). These figures are mirrored in relation to Figs. 14 and 15. Moreover, results presented in Figs. 17 and 18 are based on measurements taken for larger area than in the case of the panel side with stiffeners. In the case of the flat panel side, much larger area below the excitation point was measured. This was done in order to visualize the mode conversion better due to interaction of S_0 mode with stiffeners. Thanks to this mode conversion effect, the internal structure of the panel is visible. It should be also emphasized here that by covering the panel surface with reflective tape, the signal-to-noise ratio (SNR) is improved significantly because laser beam reflection is much stronger. This reduces the noise level, which is clearly visible in Figs. 14 and 15. Noises are visible in the regions where the guided wave have not reached yet. During the measurements, different numbers of averaging in one measurement point were utilized. In the case of measurements on the surface covered by retro-reflective tape, it was three times for 50 kHz and five times for 100 kHz. In the case of no surface preparation, it was 20 and 40 averages for 50 and 100 kHz, respectively. The time of measurements increases with the number of averaging.

The results presented in this section show that scanning laser vibrometry is a very useful tool for damage detection and localization. Guided wave propagation is a very suitable technique for the assessment of damage in composite samples. The results show that utilization of retro-reflection tape improves the results, which is, however, is not necessary. However, signals need to be averaged many times in order to improve the SNR. Surface preparation is practically not applicable for measurements in the case of real structures.

5. Assessment of Adhesive Bonds using Elastic Wave Propagation

In this section, the results of the application of guided wave propagation method for bond assessment in composite structures are presented. Measurements were performed in the case of GFRP panel with bonded L-shaped stiffener (Fig. 19). The panel had dimensions 500 mm × 500 mm × 3 mm and consisted of 12 layers of VV 192T 202 IMP503 pre-preg. The stiffener was bonded to panel using 3M AF3109 2U bonding agent. In the middle of the stiffener, a debonding for a length of 50 mm and through the whole stiffener width was introduced (Fig. 19). Guided waves were generated using piezoelectric transducer attached to the panel from the stiffener side (see Fig. 19). Signals were measured by laser vibrometer on the flat panel side. In Figs. 20(a) and 20(b), RMS energy maps for propagation of guided waves with frequencies 50 and 100 kHz are presented.

Elastic wave energy concentrates mostly near the bottom panel edge where waves were generated. Wave energy is strongly reduced

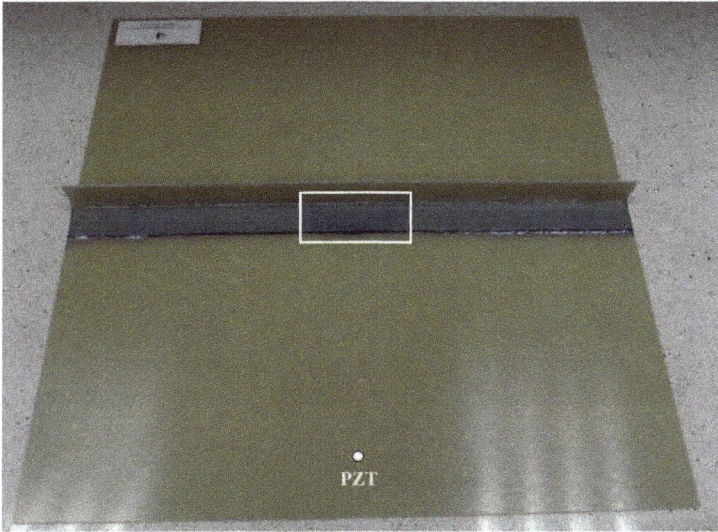

Figure 19. GFRP panel with L-shaped stiffener (damage: debonding of the stiffener).

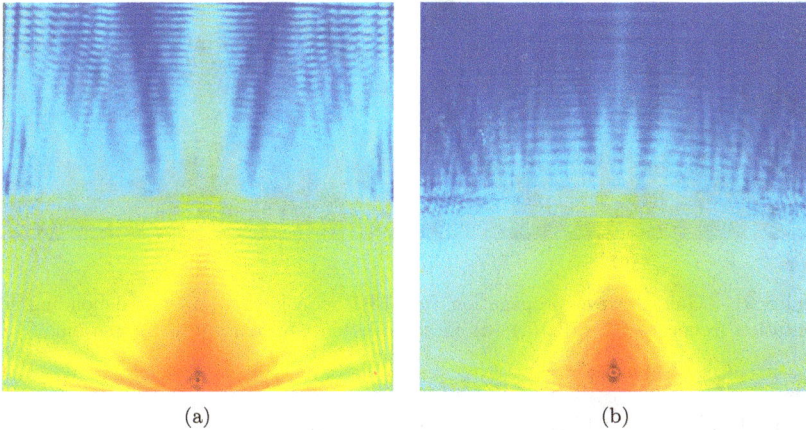

(a)   (b)

Figure 20. RMS energy maps for GFRP panel with stiffener debonding for excitation frequency: (a) 50 kHz and (b) 100 kHz (measurements taken at flat side).

due to the wave propagation across the stiffener. Moreover, energy concentration can be noticed in the location of the stiffener debonding area. In this area, waves propagate mostly in the panel. Therefore, their amplitudes are not so strongly reduced by interaction with the stiffener.

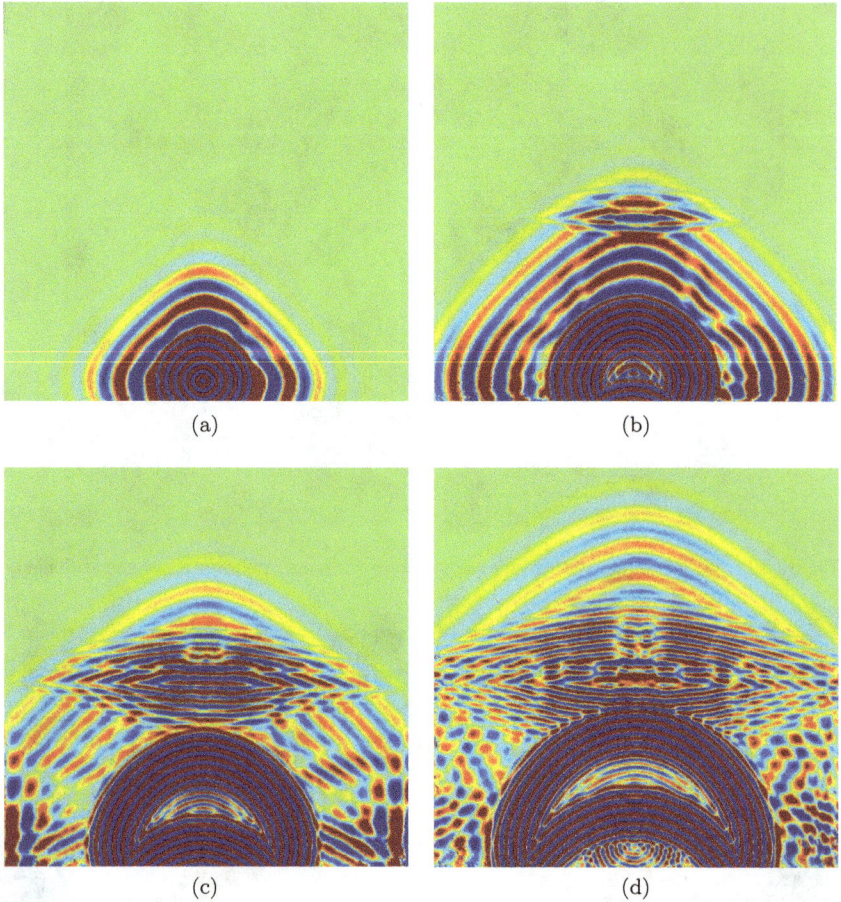

Figure 21. Guided wave propagation in GFRP panel with stiffener debonding for
excitation frequency 100 kHz: (a)–(d) chosen consecutive frames from animation.

In Figs. 21(a)–21(d), selected frames from animation of guided wave
propagation in a GFRP panel with stiffener debonding are presented. In this
case, excitation frequency was 100 kHz. In Fig. 21(a), propagation of the
fundamental mode A_0 and S_0 can be noticed. The velocity of S_0 mode is
higher in the direction of $0°$ and $90°$ due to orientation of reinforcement
($0°–90° \, 2 \times 2$ twill) in these directions. In Fig. 21(b), interaction of the S_0
mode with stiffener can be noticed. Due to this interaction, S_0 to A_0 mode
conversion occurs. Moreover, the A_0 mode (excited directly and not from

mode conversion) reflects from the bottom edge of the panel. In the frame presented in Fig. 21(c), it can be noticed that the S_0 mode propagates through the stiffener and reflects from the left and right edges of the panel. Moreover, variations in the propagation of the A_0 mode through the debonded area can be noticed. A similar situation can be observed in Fig. 21(d) where the interaction of mode A_0 with debonding can be easily distinguished.

The results presented in this section show that this method is also suitable for detection of debonding in composite structural parts.

6. Summary

This chapter presented results of research focused on the detection of degradation in composite structures. Piezoelectric sensors were used as the active elements in two methods considered: electromechanical impedance and guided wave propagation. Both methods were tested in laboratory conditions. In this research, both methods were investigated separately but one can imagine that application of these methods in real structures (such as aircraft) could be realized, for example, using the Acellent SMART Layer that is bonded to the inner surface of the fuselage. Such an approach allows for local EMI-based assessment and inspection of the larger area of the fuselage with guided waves. Guided waves excited by the sensors were registered using a laser vibrometer. Laser vibrometry is a very useful tool for damage detection and localization. Guided wave propagation is a very suitable technique for damage assessment of composite samples. The results show that the utilization of retro-reflection tape improves the results though it is not necessary. However, signals need to be averaged much more in order to improve the SNR.

Acknowledgments

The authors of this paper would like to gratefully acknowledge that this research was supported by Polish National Centre for Research and Development (NCBiR) granted by agreement number PBS1/B6/8/2012 (project KOMPNDT).

Research was also partially supported by the National Science Centre of Poland granted by agreements number: UMO-2014/13/D/ST8/03167 and UMO-2013/11/D/ST8/03355.

References

1. La Saponara, V., Environmental and chemical degradation of carbon/epoxy and structural adhesive for aerospace applications: Fickian and anomalous diffusion, Arrhenius kinetics. *Composite Structures* **93** (2011) 2180–2195.
2. Kumar, B. G., R. P. Singh, and T. Nakamura, Degradation of carbon fiber-reinforce epoxy composites by ultraviolet radiation and condensation. *Journal of Composite Materials* **36**(24) (2002) 2713–2733.
3. Brune, K., L. Lima, M. Noeske *et al.*, Pre-bond quality assurance of CFRP surfaces using optically stimulated electron emission. In *Proceeding of the 3rd International Conference of Engineering Against Failure (ICEAF III)*, Kos, 26–28 June 2013, pp. 300–307.
4. Ehrhart, B., B. Valeske, M. Sarambe, N. Chobaut, A. Gendard, and C. Bockenheimer, Preliminary tests for the development of new NDT techniques for the quality of adhesive bond assessment. In *2nd International Symposium on NDT in Aerospace*, Hamburg, Germany, 2010.
5. Markatos, D. N., K. I. Tserpes, E. Rau *et al.*, Degradation of mode-I fracture toughness of CFRP bonded joints due to moisture and release agent and moisture pre-bond contamination. *Journal of Adhesion* **90** (2014) 156–173.
6. Markatos, D. N., K. I. Tserpes, E. Rau *et al.*, The effects of manufacturing–induced and in-service related bonding quality reduction on the mode-I fracture toughness of composite bonded joints for aeronautical use. *Composites Part B: Engineering* **45** (2013) 556–564.
7. Bhalla, S., A. Gupta, S. Bansal *et al.*, Ultra low-cost adaptations of electro-mechanical impedance technique for structural health monitoring. *Journal of Intelligent Material Systems and Structures* **20** (2009) 991–999.
8. Park, G., C. R. Farrar, F. L. di Scalea, and S. Coccia, Performance assessment and validation of piezoelectric active-sensors in structural health monitoring. *Smart Materials and Structures* **15**(6) (2006) 1673–1683.
9. Moharana, S. and S. Bhalla, Influence of adhesive bond layer on power and energy transduction efficiency of piezo-impedance transducer. *Journal of Intelligent Material Systems and Structures* **26**(3) (2014) 247–259.
10. Bhalla, S. and S. Moharana, A refined shear lag model for adhesively bonded piezo-impedance transducers. *Journal of Intelligent Material Systems and Structures* **24**(1) (2013) 33–48.
11. Schwankl, M., Z. S. Khodaei, M. H. Aliabadi, and C. Weimer, Electro-mechanical impedance technique for structural health monitoring of composite panels. *Key Engineering Materials* **525–526** (2013) 569–572.
12. Park, S., C.-B. Yun, Y. Roh, *et al.*, Health monitoring of steel structures using impedance of thickness modes at PZT patches. *Journal of Smart Structures and Systems* **1**(4) (2005) 339–353.
13. http://en.wikipedia.org/wiki/Frechet_distance.
14. Efrat, A., L. J. Guibas, S. Har-Peled, J. S. B. Mitchell, and T. M. Murali, New similarity measures between polylines with applications to morphing and polygon sweeping. *Discrete and Computational Geometry* **28**(4) (2002) 535–569, doi: 10.1007/s00454-002-2886-1.

15. Malinowski, P., T. Wandowski, and W. Ostachowicz, Study on adhesive bonds influence on EMI signatures. *Structural Health Monitoring 2015: System Reliability for Verification and Implementation 1–2. Proceeding of 10ᵗʰ International Workshop on Structural Health Monitoring* Vol. 1–2, (2015), pp. 213–220.

16. Staszewski, W. J., S. Mahzan, and R. Traynor, Health monitoring of aerospace composite structures — active and passive approach. *Composites Science and Technology* **69**(11–12) (2009) 1678–1685.

17. Wilcox, P. D., Omni-directional guided wave transducer arrays for the rapid inspection of large areas of plate structures. *IEEE Transactions on Ultrasonics, Ferroelectrics, and Frequency Control* **50**(6) (2003) 699–709.

18. Yu, L. and V. Giurgiutiu, *In situ* 2-D piezoelectric wafer active sensors arrays for guides wave damage detection. *Ultrasonics* **48**(2) (2008) 117–134.

19. Giurgiutiu, V., *Structural Health Monitoring with Piezoelectric Wafer Active Sensors.* Elsevier Academic Press, 2008, p. 760, ISBN 978–0120887606.

20. Ambrozinski, L., Stepinski, T. and Uhl, T., Efficient tool for designing 2D phased arrays in lamb waves imaging of isotropic structures. *Journal of Intelligent Material Systems and Structures* **26**(17) (2015) 2283–2294.

21. Lee, S. J., N. Gandhi, J. S. Hall, J. E. Michaels, B. Xu, T. E. Michaels, and M. Ruzzene, Baseline-free guided wave imaging via adaptive source removal. *Structural Health Monitoring* **11**(4) (2012) 472–481.

22. Sharif Khodaei, Z. and M. H. Aliabadi, Assessment of delay-and-sum algorithms for damage detection in aluminium and composite plates. *Smart Materials and Structures* **23**(7) (2014) 075007.

23. Su, Z., L. Ye, and Y. Lu, Guided Lamb waves for identification of damage in composite structures: A review. *Journal of Sound and Vibration* **295**(3–5) (2006) 753–780.

24. Ruzzene, M., Frequency-wavenumber domain filtering for improved damage visualization. *Smart Material and Structures* **16**(6) (2007) 2116–2129.

25. Carrara, M. and M. Ruzzene, Frequency-wavenumber design of spiral macro fiber composite directional actuators. *SPIE Smart Structures and Materials* 94350M-12 (2015).

26. Kudela, P., M. Radzienski, and W. Ostachowicz, Identification of cracks in thin-walled structures by means of wavenumber filtering. *Mechanical Systems and Signal Processing* **50–51** (2015) 456–466.

27. Wandowski, T., P. Malinowski, J. Moll, M. Radzienski, and W. Ostachowicz, Analysis of guided wave propagation in a tapered composite panel. *Proceedings of SPIE 9438, Health Monitoring of Structural and Biological Systems* (2015), 94380Q, doi: 10.1117/12.2083760.

28. Moll, J., T. Wandowski, P. Malinowski, M. Radzienski, S. Opoka, and W. Ostachowicz, Experimental analysis and prediction of antisymmetric wave motion in a tapered anisotropic waveguide. *Journal of Acoustical Society of America* **138**(1) (2015) 299, doi: 10.1121/1.4922823.

Chapter 5

Design and Development of a Phased Array System for Damage Detection in Structures

Bruno Rocha, Carlos Silva, Mehmet Yildiz and Afzal Suleman*

University of Victoria, Department of Mechanical Engineering, Victoria, BC, V8W 3P6, Canada
suleman@uvic.ca

The development of a structural health monitoring (SHM) strategy based on a piezoelectric (PZT) phased array system is presented. The objective is to increase the low signal-to-noise ratio (SNR) compared to PZT networks for Lamb wave-based SHM systems. This is achieved by constructive interference, known as beamforming, of different waves generated by transducers in the array. By carefully selecting and changing the delays between actuation of consecutive transducers in the array, the wavefront can be steered to a range of selected directions in the plane of the array. The system is based on the fast propagating first symmetric Lamb wave mode (S0). The accuracy of the method is strongly dependent on a precise multiple actuation system and particularly the accuracy at which the diminutive time delays are introduced between actuation of the different array elements. This problem was addressed by designing and implementing a custom-made multiple actuation system. Increasing the amplitude of generated waves coupled with a high SNR improves damage detection. Damages were simulated by surface and through-the-thickness holes and cuts with different orientations and sizes. Tests were performed on both metal and composite plates, subjected to different boundary conditions, and a successful and repeatable detection of 1 mm cumulative applied damage was realized.

1. Introduction

Inspections for the assessment of structural condition are of the utmost importance for a safe and efficient operation of aircraft in use. Aircraft structures operate in harsh conditions, sustaining high loads, fatigue cycles and extreme temperature swings. The characteristics of aircraft operation, at altitude and high speeds both in air and on the ground, at take-off and landing, usually lead to catastrophic consequences in terms of loss of life and economic loss when failure of a primary aircraft structure occurs. To achieve much needed lighter structures, aircraft structures are designed following a damage-tolerant philosophy. Also to reduce structural weight, newer materials, such as composites, have been considered. These present radically different characteristics than those extensively known from isotropic materials widely used in aircraft (e.g., aluminum), for instance in terms of response to damage existence or stress concentrations. Furthermore, current aircraft fleets are rapidly ageing, while air travel is considerably increasing, with aircraft structures being introduced in operation presenting increased capacity and complexity. These characteristics are driving current research to increase reliability and simplify the application of structural inspections.

Non-destructive testing and evaluation (NDTE) techniques developed in the last decades and currently applied to assess the health of aircraft structures in operation are external to the aircraft and are confined to localized damage detection. These techniques are capable of detecting structural damage only around their application region, which requires the repetitive execution of inspections in different areas of the structure, with necessary direct access to structural areas to be inspected. This involves complicated, time-consuming and consequently expensive disassembling and assembling operations, forcing prolonged grounding of the aircraft. Inspection and maintenance procedures are costly to operators and are performed on the basis of pre-drawn schedules, even as the pressure from aircraft fleet operators is mounting to reduce inspection and maintenance time and extend intervals between consecutive interventions. This increases risk of existence of undetected damage and its progress between inspection operations, for instance due to unpredicted flight severity or foreign object impact. Also, damage might exist and grow in regions that are not inspected.

Structural inspections may also be performed without any damage being detected, incurring costs that could have been avoided. Structural components are sometimes replaced without having reached their operational limit, since it is less expensive to replace components approaching their predicted operational lifetime immediately while the structure is accessible during a scheduled inspection. Otherwise, if the component is maintained during operation, the consecutive maintenance interval would have to be reduced considerably. Moreover, maintenance operations also involve certain risks of potentially creating additional damage to the aircraft structure, without anyone being aware of it. These potential damages add to possible manufacture imperfections, impact damage, corrosion, fatigue-related damage for instance due to pressurization cycles, excessive loads, unavoidable stress concentration locations in design, etc.

Currently, NDTE techniques do not provide for persistent, integrated, real-time and global structural integrity evaluation systems. The drawbacks in the current application of NDTE techniques to aircraft structures in operation are responsible for the development of structural health monitoring (SHM) methods. SHM techniques are currently being intensively researched, with the first significant developments emerging at the beginning of the last decade, enabled by advances in electronics and computation. These systems are intended to be embedded into the structure to be monitored, assessing its structural condition in real time or near real time, in a more global way. Inspections can be performed during the flight in operation, without requiring direct access to the structure, with no operation downtime and without applying long and expensive disassembling and assembling procedures. Then, condition-based maintenance (CBM) could also be performed.

Reliable SHM systems can potentially identify damage in its earlier stages of development, particularly if systems allow for a reduction of the minimum damage dimension to be reliably detected. Design safety factors could be decreased, particularly in areas of stress concentration and those prone to fatigue, as for instance in riveted joints, or composite material joints, where design safety factors are considerably increased and structural reinforcement is required. The decrease of design structural safety factors, without deterioration in the safety of operation, and subsequent reduction of required structural reinforcements and structural weight will have a direct impact on increasing available payload weight. Alternatively, the reduced weight will require a reduced lift, reducing generated drag, required thrust and importantly fuel consumption (with evident economic benefits) and emissions (achieving a more environmentally friendly aircraft), eventually reducing the weight of the aircraft even more. Since aircraft design is a cyclic process, small weight reductions in the beginning of the process are greatly amplified during the design cycle.

Similar to what is applied presently with NDTE in aircraft structures, knowing the predetermined flight severity (flight loads spectrum) and having in mind that predictive algorithms for damage growth are established (damage prognosis), the remaining useful life (RUL) of the component could be estimated, but now in real time. Future aircraft operation could be tailored and better fleet management can be achieved. The information provided by SHM systems is also fundamental for maintaining legacy aircraft in operation, for life extensions and when aircraft assumes different roles from the ones they were designed for.

Several SHM methodologies are being actively researched and developed with the objective of detecting, locating and characterizing damage such as identifying damage type, shape, dimensions, orientation, etc. From these methods, piezoelectric (PZT)-based phased arrays that generate and sense Lamb waves have shown promising results, being capable of detecting small defects. The SHM systems based on phased arrays achieve improved results through the amplification of the interrogation waves and consequently of sensed reflected waves, including potential damage reflections. This enables an improved signal-to-noise ratio (SNR). Additionally, phased array-based SHM systems are a natural evolution from ultrasound-based NDTE systems, widely used in aircraft structures in operation and considered a valuable inspection technique, with an available broad and valid knowledge base. Lamb wave based SHM systems do not require any special precautions in their application, for operator safety or to avoid damage to the surrounding systems, as Eddy currents-based methods do with possible electromagnetic interference (EMI) to surrounding systems.

The developed phased array systems presented here are based on the excitation and sensing of the fast propagating first symmetric Lamb wave mode (S_0) waves and wavefronts and resultant reflected S_0 waves (at boundaries, reinforcements, discontinuities and potential damages). The decision to develop a system based on the S_0 Lamb wave mode emerged from the high propagation velocity of such waves. These waves and their subsequent reflections generated by potential defects are then less prone to interference from slower propagating wave modes. These waves are also more sensitive to internal structural damage, while presenting lower amplitude damping during propagation. This aspect means that they are able to propagate through larger areas, increasing the inspection region. However, their high propagation velocity has the disadvantage of imposing restrictive and demanding requirements regarding the development of actuation and data acquisition systems, considering also their dispersive behavior. This is in fact the reason for the relatively small number of reports in the literature on the successful development and application of SHM systems based on the S_0 wave mode, particularly considering dedicated phased array systems. PZT transducers are selected due to their capability of simultaneous high frequency actuation and sensing. Effort was directed at minimizing the number of transducers to be applied and a linear phased array was implemented as the basis for the future development of arrays based on more complex shapes, such as cross, star and circular shapes. The research described in this chapter comprises a study of Lamb waves and their propagation characteristics based on the host material to be inspected and applied to the implementation of phased array SHM systems. This study led to the selection of transducers to be employed and to the development of dedicated actuation, signal acquisition and data processing subsystems, and damage detection and location algorithms.

Tests for damage detection and location were performed on metal and composite plates, representative of aircraft panels, in laboratory and aircraft maintenance workshop environments, with no surrounding noise control and limited temperature control. Different boundary conditions were applied to the panels, from fully supported to simply supported, and riveted boundary conditions were assessed. Damages were inflicted cumulatively to the plates, being simulated by the thickness and surface holes and cuts with different orientations. Defects, as small as 1 mm, were successfully detected and their position was determined in 80% of the inspections performed. One of the main objectives accomplished in this work was to achieve the reliable detection of smaller damage in comparison to what is currently the standard for NDTE techniques applied to aircraft structures in operation, independently from prescribed boundary conditions.

2. State of the Art

A phased array consists on applying a certain number of transducers, which may be aligned in various configurations, depending on the application (in linear, round, cross, star, diamond fashions, etc.). By sequential activation, with a predetermined time delay introduced in between actuation of neighboring transducers, beam forming is possible through constructive interference of different propagating Lamb waves, generated by individual actuators. Through the variation of such time delay, the wavefront can be emitted according to different predefined directions (steered). The time delays for a certain desired direction are determined knowing the wave propagation velocity of interest. Another possibility is to focus the wavefront in a certain point (or region around a pre-determined point). For instance, in a damage detection scheme applying a phased array, first a scan is executed in a certain direction (with the wavefront emitting and propagating in that direction). If the detection of a possible damage reflection occurs and it is suspected that damage exists in a certain direction, the location of damage is calculated through time-of-flight (ToF) of the reflected wave. Afterwards, the wavefront can be focused at that point, increasing the amplitude of incident beam even more and consequently damage reflection. This can be achieved by performing a single scan, introducing different time delays in between the actuation of consecutive transducers.

Similarly, the phased array principle can be applied to sensing, i.e., wave sensing and potential damage-generated reflections sensing capabilities can be focused into a certain direction. This can be achieved by gathering the sensed signals from different transducers after the execution of a scan (with an actuation either introduced by phased array or by a single actuator, or even with an external actuation) and shifting neighboring sensor signals by a certain time delay. This time delay is equal to the difference in times at which an incoming wave from the selected direction would reach two consecutive transducers in the array. With this procedure, the reflection from a potential damage existing in that particular direction will appear in all sensor signals at the same time. Afterwards, the different shifted sensor signals are added, so that the potential damage reflection in the scanned direction will be enhanced with relation to the remaining sensor signal. Particularly, the detected reflection will be enhanced with relation to noise and other reflections, for instance from other damages in other directions, or boundaries. This is equivalent to assuming a potential damage reflection as a beam, travelling through the different sensors. This approximation potentially enhances damage detection capabilities, since the resultant augmented reflected wave corresponding signal is likely to be detected. However, errors are introduced in the determination of damage location by assuming that the reflected wave is rectilinear (as a beam) and not curved as it is in reality. These errors are smaller if damage is distant from the phased array, since the damage reflected wave segment passing by the phased array will be more approximately rectilinear.

The first advantage of a phased array actuation, in relation to a transducer network configuration, is the amplification of the original mechanical deformation wave actuation. Such actuation enhancement is not based on electric or electronic amplification, which can be, however, optionally implemented. Usually electric or electronic amplification also introduces additional noise in actuation and consequently in wave pattern frequency and propagation velocities, and subsequent sensing (possibly generating unwanted waves). Through amplification of actuated wavefront, potential damage reflections will consequently present increased amplitude, increasing the likelihood of being detected. Detectable damage size might also be reduced. Even more importantly, the difficulties created due to damping of propagating waves, particularly when structural reinforcements exist, can be diminished.

A second advantage is the capability of scanning the component in all directions, through the execution of several scans in different directions, with the introduction of different time delays. With relation to transducer networks, phased arrays present the advantage of being capable of steering the generated wavefront, i.e., to emit the enhanced actuation in a determined direction, focusing inspection effort towards a certain predetermined area.

One disadvantage of phased arrays is that the activated wavefront originates always from the same point, or region, in the component under inspection. With the application of transducer networks, in different scans, different transducers in different positions can be actuated. In the phased array approach, since the generated wave origin is always in the same location, the incident wave with relation to a potential damage will always come from the same direction and side, which is not true for transducer networks. As a result, damages that present a detectable dimension at least in one direction, even if their dimensions in other directions are considerably small, are prone to be detected with a network. This might not be true for phased arrays. A damage that presents a small dimension perpendicular to a radial direction with relation to the phased array (considering the origin as the phased array position) might not be detected even if it has considerable and detectable dimensions along that radial direction. This disadvantage is, however, diminished by increased amplitudes introduced by the phased array, either in actuation and generated wavefront, or in the resulting damage reflection corresponding to sensed signal.

The principle applied to phased arrays is similar to what is applied in static radar arrays, medical imaging ultrasounds and conventional NDTE ultrasonic inspection phased array systems.[1] The difficulty in the application of phased arrays to SHM, involving Lamb waves and their generation, is mainly related with the required phased actuation system. Due to high propagation velocities of Lamb waves, such system must be capable of reliably and accurately introducing diminutive time delays involved in the phased array approach. Simultaneously, all the requirements related to Lamb wave generation must be considered. Particularly more complex generation signals are involved with the required increased amplitude, time and specifically frequency definition. Such accuracy is even more important when the fast propagating S_0 wave is selected as the mode of interest to be activated by the phased array and to base the damage detection system on. For these reasons, the development of automated and dedicated phased array actuation systems for S_0 Lamb wavefront emission is not widely reported in literature.

Actuation systems for the applications referred at the beginning of the previous paragraph rely to a large extent on the generation of bulk pulses or pulse trains. In the specific case of conventional NDTE ultrasonic inspections, based on phased array systems involving high frequencies, such actuation waveforms degenerate in the actuation of the scattered bulk body waves. Application of a phased array reduces the difficulties created by such scattering, by promoting a wavefront with relevant amplitude, propagating into a determined direction of inspection, i.e., focusing the inspection efforts into a determined region, even though those systems are only capable of localized damage detection and particularly through the thickness of the structure to be inspected (due to significant wave scattering, poor definition and damping of propagation amplitude).

A phased array actuation system cannot be based, at least uniquely, on simple multiplexing techniques involving a single-channel signal generation input and multiple outputs — one for each actuator. With such approach, the application of minimal time delays — corresponding to the generation of a wavefront propagating in directions approaching the perpendicular to the phased array — would not be possible. The reason is that with such small delays, although phased, actuation will occur in different transducers at a certain time. To consider multiplexing, delay or repeater circuits must be employed in each channel output, for each actuator, usually at the cost of losing signal definition. Without taking into account multiplexing, a dedicated signal generation and actuation circuit must be developed and applied to each actuator. A precise master circuit must then be implemented to control time delays and the activation of each slave actuator circuit.

The application of multiple actuation circuits results in complex and costly systems. Currently, there are reported efforts in research, development and application of this type of systems for both conventional NDTE ultrasonic inspection using phased arrays and SHM systems.

Accellent[a] developed a similar system that is now commercially available. However, besides elevated cost, it does not allow a simple selection of generation signal waveform. At the same time, this system requires application of a considerable number of transducers to achieve reliable damage detection. Furthermore, minimum damage-detectable dimensions are in the order of centimeters. There are also reported efforts to adapt National Instruments[b] boards for signal generation in multiple channels. It was verified that synchronizing time accuracy of the master system, necessary to apply the required precise time delays, was insufficient.

Due to difficulties related to actuation of phased array Lamb wave-based SHM systems, the majority of reported research uniquely involves the referred phased array principles in sensing. Bao[2] and Purekar[3] studied this approach applied to linear arrays by means of simulation and experimental work, successfully detected and located a 19-mm crack introduced in a 2-mm thick aluminum plate. Pena *et al.*[4] and Malinowski *et al.*[5] reported the successful application of SHM Lamb wave-based phased arrays for damage detection. These systems also included dedicated actuation systems developed by the authors. To decrease the complexity of such actuation systems, the authors selected to base the system and damage detection algorithm on the slower A_0 Lamb wave mode.

Salas and Cesnik[6] developed a new concept of PZT transducer fabrication meant to produce similar results as a phased array. In this particular case, one probe contained eight independent transducers, each responsible for generating a determined azimuth. Analytical studies and experimental damage detection scans, with the introduction of simulated damages, were performed successfully. The main advantages reported were its flexibility and conformability to curved shapes, its capability of generating multiple modes and the possibility of independent sensor/actuator function for each one of its individual transducers.

In relation to signal processing techniques, directly from sensor signals, in amplitude versus time, it is possible to extract corresponding wave mode signals, their boundary reflections, ToF and consequently propagation velocities. To improve the accuracy of the assessment of sensor data, by reducing the influence of noise, data post-processing techniques must be applied. For instance, Bao[2] used the signal root-mean-square (RMS) and energy densities by Hilbert transform towards this

---

[a] Accelent Technologies, http://www.acellent.com, last accessed on August 30, 2010.
[b] National Instruments, http://www.ni.com, last accessed on August 30, 2010.

objective. However, further data processing techniques must be applied to achieve an accurate damage detection based on the comparison between a baseline (corresponding reference or undamaged state signal) and the current or damaged state signal. Usually, normalization and data shifting in time (tuning/aligning time origins) are necessary to synchronize both the responses. Michaels and Michaels[7] used both normalization and data shifting during a damage detection experiment on an aluminum plate with a bonded transducer network of four elements. Using this approach and with damages introduced near sensors, 6.4 mm holes were successfully detected.

Time reversal can also be used for damage location. The implementation of this technique does not require previous knowledge of a baseline response for future comparison with inspection scan responses. However, the propagation velocity of the waves (and wave mode) of interest must be known. Time reversal is particularly interesting for passive sensing applications, i.e., when an actuation is not required. For instance, time reversal can be applied in the implementation of acoustic emission detection methods.[8] Sohn et al.[9] also applied this method successfully for the detection of delamination on a composite panel, with a combined transducer actuation. On the delaminated plate, the sensed signals mismatched the original signals created by the actuator. This approach was focused on assessing the mentioned feature, only for waves propagating in the direct path established between the sensor and the actuator. With this limitation, time reversal signal processing techniques became feasible only when relatively high-density transducer networks were applied or in 1D specimens. Nevertheless, very small damages could be detected.

Xu and Giurgiutiu[10] dedicated part of their research to assess the effects of mode selection, i.e., actuation wave tuning, on the application of time reversal post-processing techniques. They concluded that system tuning for actuation (and posterior sensing) of a single-wave mode is essential to enhance the reliability of the method.

Post-processing techniques commonly considered for Lamb wave-based SHM-derived data include Fourier transforms (FTs), fast Fourier transforms (FFTs) and 2D-FFT. These techniques allow for posterior signal filtering, signal reconstruction, noise reduction and ultimately to focus on damage detection algorithms in the frequency of interest (of actuation). The last aspect is particularly interesting if care is not dedicated to achieving an accurate and efficient actuation, centered in a single frequency, with the consequent generation of multiple groups of waves with different frequencies and propagation velocities. In that case, these techniques enable the selection of the waves (frequencies) to be considered in the SHM inspection system. Interesting experimental results were reported by Gomez-Ullate et al.[11] using 2D-FFTs on data collected using a vibrometer. Gao et al.[12] performed similar experiments, with the implementation of referred techniques, using laser actuation and sensing, on a copper plate. Based on identified waves and corresponding modes and properties, the plate's mechanical properties were calculated.

Joint time and frequency domain analysis can also be applied to sensor signals for data post-processing. These methods conjugate both phase and frequency variations along the time response. For such analysis, sensor signals (amplitude versus time) are divided into different time segments. In a specific time range, the signal is then decomposed into several amplitude and phase components for different frequencies. These methods promote a more accurate signal assessment in the time segments of interest, by either noise reduction, and/or allowing to assess uniquely one or more specific frequencies of interest. Two interesting techniques included in this particular type of data post-processing are short-time Fourier transform (STFT) and wavelet transform (WT). Paget et al.[13] executed experimental impact tests on a composite panel, with the application of WTs to decompose sensor data. Relative WT coefficients changed accordingly from a healthy condition to three different conditions of increasing levels of impact energy. This approach provided a significant information on damage assessment.

In order to assess damage severity, damage indices are also usually applied. Damage indexes are established from the comparison of certain features, extracted from sensor signals. One interesting damage index is based on distance to amplitude correlation (DAC) of the propagating waves, similar to what is used in conventional NDTE ultrasonic inspections. DACs are established based on propagation wave amplitude damping. This can also contribute to the determination of the distance between the sensors and the damage. Damage indices are based on the correspondence between the amplitude of the damage-generated reflections and projected damage dimensions in a perpendicular direction to wave propagation, promoting the determination of damage severity. Basically, after damage detection and determination of damage location, and distance between the damage and the sensor, DAC can be applied to obtain damage reflection amplitude, at damage position, from the sensed reflections amplitude. Based on the corrected sensed reflections amplitude, through DAC, damage indexes can be established. Examples of the application of this feature are given by applying it to signal amplitude in time domain,[14] to RMS magnitude in the time domain,[15] or in frequency domain[16] and combined domains.[17]

After signal acquisition and data post-processing, qualitative and/or quantitative characteristic features related to damage existence can be extracted. Depending on transducers' disposition (networks or arrays) and on extracted features, different algorithms for damage location can be applied.

The application of neural networks (NNs)[18] has also been and is still being actively investigated for fast and accurate data post-processing in Lamb wave-based SHM systems, with promising results. The implementation of neural networks is based on the consideration of different features of sensors' signals and optionally on the different sensors themselves. Neural networks are based on multiple, individual decision-making neurons, which must be trained before to achieve a correct decision output. This is in fact the main disadvantage of this technique: it is not readily available. Long learning periods and complex training are required. This technique can learn from previous results obtained from the system (and structure, geometry, material, etc.) to which it is going to be applied. Such results might be obtained either experimentally or through system simulations. Considering damage detection systems, it is not practical in terms of time and cost to execute multiple experiments, considering multiple damage types, positions, orientations, etc. For instance, if it is desirable to test one component with a single damage, varying damage dimensions, or orientation, numerous test components would have to be damaged, one for each experiment to be performed. The obvious solution is that NN must learn from simulation results. This justifies the development of numerical simulations discussed before. However, simulations are still not perfect and do not absolutely match reality. Difficulties in damage simulation, wave propagation pattern simulation, among others, still exist.

When transducer networks are considered, there are two possible approaches for damage detection: pitch-catch and pulse-echo. Pitch-catch is based on the detection of the scattered wave caused by existence of the damage in the direct path between an actuator and sensor pair. A relative high number of sensors is frequently needed in this case. The simplest approach implies creating a sufficient dense grid, established by numerous direct paths established between every transducer pair available. Ihn and Chang[19] tested this principle successfully on an Airbus aircraft panel. Using two strips with 18 PZT transducers each, damage indexes based on sensed wave energy were calculated. Cracks with dimensions as small as 4 mm were detected, with damage length being linearly proportional to established damage indexes.

The application of time reversal techniques is also adequate for pitch-catch implementations. The intersection of at least two direct concurrent paths established by different pairs of transducers, whose reversibility failed, indicates damage location. The relationship between reversibility magnitudes detected can contribute to the determination of damage dimensions.

Pulse-echo techniques rely on the fact that activated wave echoes produced by a damage can be captured by sensors in the network. This is an active approach requiring actuation and activation of waves. It totally depends on successful and reliable detection and assessment of potential damage-reflected waves in corresponding sensed signals. A smaller number of sensors is required, in comparison with the previous approach. Disadvantages of this technique are also related to the smaller number of sensors employed and the possibility of active waves having to travel longer distances, with considerable amplitude attenuation, to reach the sensors. If that occurs, the method will present inferior detectability and will be more prone to noise interference, with consequent information loss. Raghavan and Cesnik[20] performed tests for damage location using a transducer network. Holes with diameters as small as 5 mm were successfully detected and located applying the pulse-echo technique.

For both pitch-catch and pulse-echo approaches, the ToF of the different waves is frequently used to determine the location of potential defects, if the wave propagation velocity of interest is known. ToF consists in the time interval between the successful detection of a (potential damage generated) wave reflection, in the corresponding sensor signal, and actuation, i.e., the propagation time of the original activated wave plus the propagation time of a (potential damage) wave reflection. For an accurate determination of a potential damage location, this time must be precisely established. For the particular case of pulse-echo, the application of at least three sensors enables the use of triangulation schemes.

Another available method is named migration technique. This approach does not require a baseline for comparison and is based on a geophysical method to detect seismic epicenters, in use for the last 50 years. Some research efforts have been dedicated to exploring the viability of application of such method in Lamb wave based SHM. Wang and Yuang[21] studied its implementation to the inspection of a quasi-isotropic composite panel with two distinct delaminations. The active wave propagation pattern was calculated by time reversal of the sensed response in several time steps. This resulted in the development of a step by step visualization technique. In this particular case, two 0.1 m × 0.1 m delaminations were successfully identified. Here, existing damages were treated as wave sources, according to Huygens' principle.

For phased arrays, the pulse-echo technique is commonly applied. The main difference, with relation to the application of pulse-echo techniques with transducers networks, is based on the fact that the damage location azimuth is pre-determined by the phased array wavefront steering. Thus, if the echo ToF is obtained for a determined scanned azimuth, knowing the wave propagation velocity of interest, damage location can be found.

Different phased array arrangements have also been studied. Malinowski et al.[5] conducted a numerical study, with the use of spectral elements, to assess a star-shaped array performance on damage detection, with encouraging results. This shape was considered as an attempt to overcome the difficulties related to the application of linear phased arrays. Interference between transducers and generated waves from neighboring transducers (or in the sensing mode, from transducers to the damage-generated reflection wave), results in wavefront amplitude damping when the wavefront is steered along the linear phased array direction. Smaller wavefront amplitudes result in smaller potential damage wave reflection amplitudes, are more prone to noise influence and more difficult to detect in sensor signals, thus leading to a decreased damage detectability. The advantage of phased array inspections is compromised in those directions. Furthermore, linear arrays, if placed in the middle of a panel, generate two wavefronts propagating in opposite directions, such that if a damage reflection is detected there will be two possible damage positions.

Finally, Yu et al.[22] summarized analytical and experimental work using PZT transducers for damage location in thin walled structures, both in isotropic and orthotropic (composites) host mediums. The results show that the available techniques (networks and phased arrays) using different approaches (pulse-echo, pitch-catch and time reversal) are suitable for damage detection. Influence of variations in operational and environmental characteristics and reliability of bonding materials were also addressed. One of the conclusions is the required improvement in the reliability of bonding of PZT transducers. Finally, the advantages of having transducers equipped with processing wireless units were also focused upon.

In relation to phased arrays, there are no reports on the development of low-cost and dedicated automatic systems for Lamb wave-based SHM, involving accurate generation and application of fast propagating S_0 wavefronts. Previously several characteristics of the application of the S_0 mode, in comparison to the application of the A_0 mode, were referred. The former presents smaller propagation amplitude damping, while the latter presents a smaller wavelength (for the same frequency), being theoretically more capable of detecting smaller damages. However, the deformation pattern imposed by the A_0 mode (as a transverse or bending wave) results in lesser interference with damages and consequently increased difficulty in damage detection, particularly when interior damages are considered.

A definition is necessary to set the small time delays precisely in between actuation of consecutive transducers in the array, to correctly generate a wavefront. However, one of the advantages of using the S_0 Lamb wave mode, in detriment of the A_0 mode, lies exactly in its high propagation velocity, being less prone to interference from the slower A_0 mode and its reflections. Conversely, the slower propagating A_0 wave mode always appears after the S_0 mode, with the wave pattern of the former being prone to interference by the later, in particular by S_0 reflections, either damage or boundary originated. For these reasons, it was decided to develop the phased array SHM system based on the application of the S_0 Lamb wave mode.

The objective of this research is to obtain a relatively simple and low-cost system, capable of a near-term reliable application in real aircraft structural components. The final system should be competitive for replacement of the conventional NDTE method. Furthermore, such system will aim to be capable of detecting smaller damages in components with any boundary conditions, using a minimized number of sensors and with a wide inspection area coverage. Finally, the developed system will include phased arrays with a dedicated automatic system for the activation and application of S_0 wavefronts for accurate damage detection.

2.1. *Design of a Phased Array Actuation System*

One solution to increase amplitude levels of actuation and subsequently of potential damage reflection waves, particularly when it is desired to inspect small damage sizes, is to use multiple actuations. The objective is to promote constructive interference between the different generated waves in predetermined positions of the component to inspect — beam forming, generating a wavefront. Actuators will have to be in different positions on the component and a phased actuation is necessary, with small time delays between the actuation of neighboring transducers. These time delays are related with the relative positions of transducers, or the configuration of the array, the distances between transducers, considering a linear array and its pitch, the propagation velocity of waves to excite, and the desired propagation direction of the generated wavefront. By varying the time delays, the wavefront propagation direction can be modified (steered). Inspection effort can then be focused on the pre-determined regions of the component, while by modifying time delays between different scans, the component can be entirely inspected.

One of the main difficulties in the application of a phased array for SHM, based on the fast propagating S_0 wave mode, is the precise implementation of the necessary small time delays between consecutive actuations of the elements of the array. This small time delays must be applied accurately to guarantee the correct generation of the fast propagating S_0 wavefront. To address this difficulty, a dedicated multiple channel actuation system was developed to excite the PZT transducers in a linear array. A linear array configuration is preferred, since other configurations, such as cross, star or diamond configurations, can be seen as a combination of multiple linear arrays.

The phased array system was developed for application to isotropic and orthotropic materials, such as aluminum and composites, respectively. The developed system accounts therefore for different wave propagation velocities for different directions within the component to inspect. The different velocities can be set in the system by its user and/or verified experimentally in the component to inspect automatically by the system. In the last case, the system interrogates the component in different directions and the boundary reflections are searched for in the receiving sensor signals. Knowing the relative distance of the array to the different boundaries and the ToF of the corresponding identified reflections, wave propagation velocities are determined. This, in fact, became an important feature even for the experiments executed in aluminum panels, since, due to their manufacturing processes, mechanical properties are not exactly the same in all directions. Small relative differences in properties exist, for instance due to the lamination direction in aluminum plates. These small relative differences are translated to considerable absolute values when the high values of Young's modulus and wave propagation velocities are considered: differences around 200 m/s (however less than 0.5%) were verified in the propagation velocity of the S_0 wave in aluminum plates, in different directions.

2.2. Dispersion Curves

The experiments were performed on Al2024-T3 aluminum plates with dimensions of $1.5\,\text{m} \times 1.5\,\text{m} \times 2\,\text{mm}$. The dispersion curves for this aluminum plate were obtained by solving the Rayleigh–Lamb equations. With C_P and C_S being the longitudinal and transverse wave propagation velocities on the material, given by

$$C_P = \sqrt{\frac{E(1-v)}{\rho(1+v)(1-2v)}}, \tag{1}$$

$$C_S = \sqrt{\frac{E}{2\rho(1+v)}} \tag{2}$$

with E being the Young's modulus for the material, v the Poisson's coefficient and ρ the density. From C_P and C_S, the following coefficients can be determined:

$$\xi = \frac{C_S}{C_P}, \tag{3}$$

$$\zeta = \frac{C_S}{C_L}, \tag{4}$$

$$\bar{d} = \frac{\omega}{C_S} h, \tag{5}$$

where ω is the circular frequency and h is half of the thickness of the plate. From these coefficients, the Rayleigh–Lamb equations can be expressed as

$$\frac{\tan\left(\sqrt{1-\zeta^2}\bar{d}\right)}{\tan\left(\sqrt{\xi^2-\zeta^2}\bar{d}\right)} = 4\zeta^2\sqrt{1-\zeta^2}\frac{\sqrt{\xi^2-\zeta^2}}{(2\zeta^2-1)^2}, \tag{6}$$

$$\frac{\tan\left(\sqrt{1-\zeta^2}\bar{d}\right)}{\tan\left(\sqrt{\xi^2-\zeta^2}\bar{d}\right)} = \frac{(2\zeta^2-1)^2}{4\zeta^2\sqrt{1-\zeta^2}\sqrt{\xi^2-\zeta^2}}. \tag{7}$$

The phase dispersion curves obtained from Eqs. (1) to (7) for the aluminum plate were calculated and are presented in Fig. 1.

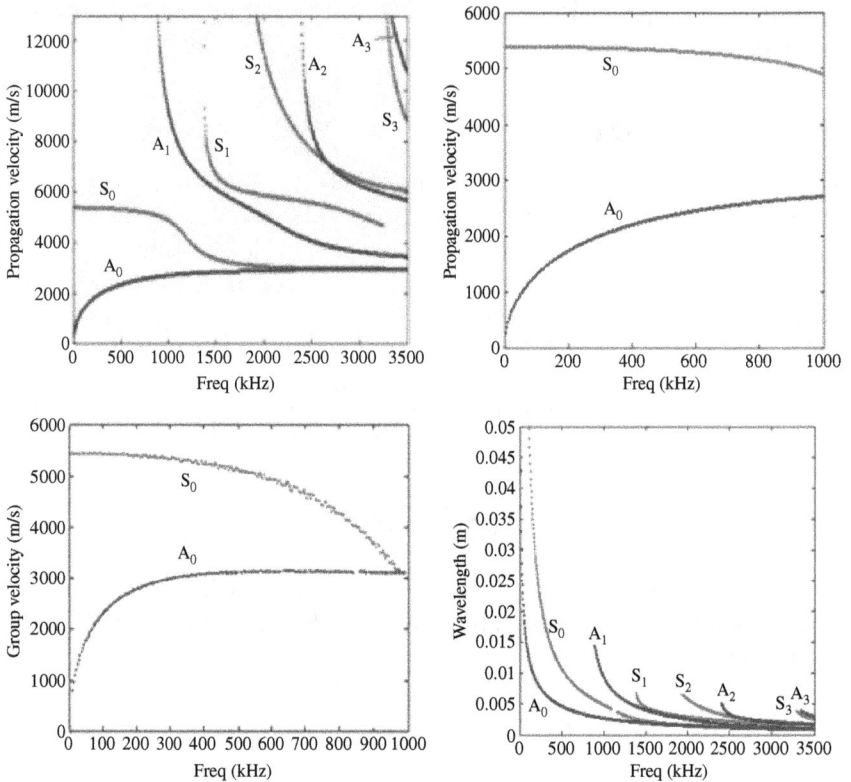

Figure 1.   Phase (top) and group dispersion (bottom left) curves, and wavelength versus frequency (bottom right) curves.

It is desirable that a single actuation will generate the smallest number of modes possible, so that the inspection effort can be focused in the use of the S_0 waves, without the generation of other modes, which will be regarded as introducing noise in the process. With this objective in mind, the actuation frequencies below 750 kHz are initially selected, since within such frequency range, only S_0 and A_0 waves are generated as observed. Furthermore, it is desirable to select an actuation frequency range for which propagation velocities are fairly constant, to minimize the dispersive behavior and scattering of generated waves, considering small imprecisions in the application of the actuation frequency. Also, since S_0 and A_0 waves are generated, it is desirable to have the maximum separation between generated waves. This translates into selecting a frequency range in which the difference between the propagation velocities of the two modes is considerable, so that S_0 waves will propagate much faster than A_0 waves, with the latter ones affecting less the acquired sensor signals, being regarded as noise to the method to be applied. From Fig. 1, a frequency range up to 400 kHz is selected.

Regarding the implementation of phased arrays, so that undesirable side lobes are not generated, the phased array pitch should be less than or equal to half of the wavelength to be excited:[23]

$$\text{Pitch} \leq \frac{\lambda}{2}. \tag{8}$$

Since the array pitch is equal to the dimension of one of its PZT transducers plus the spacing between consecutive elements in the array, the tuning of such system, i.e., the selection of the excitation frequency and wavelength, should be performed with respect to the array pitch and not just with respect to the dimensions of the PZT element. However, it is still true that the amplitude of generated Lamb waves is increased when the dimension of the applied PZTs approaches half of the wavelength of excited waves. Therefore, it was decided to implement an array with a pitch equal to half of the wavelength of waves to be activated and the PZT transducer dimensions should be as close as possible to the phased array pitch, reducing the spacing between consecutive elements to a minimum. Nonetheless, neighboring elements must not be in contact, including their bonding material, to avoid transmission between transducers, creating actuation noise and interference. A spacing of 3 mm was selected.

2.3. *Tuned Lamb Waves: Mode, Frequency and Transducer Selection*

The application of a tuned approach requires the use of transducers that will maximize the generation of the intended waves, in our case S_0 waves, while minimizing the generation of remaining modes, in our case A_0 waves. As explained previously, the pitch of the array plays a fundamental role. To achieve Lamb wave tuning, Fig. 2 is deemed essential. Specifically, tuning the system for the S_0 wave can be achieved by selecting an activation frequency and corresponding wavelength, and an array pitch equal to half of the S_0 wavelength to be excited and simultaneously a PZT transducer diameter multiple of the A_0 wavelength to be generated. To refer that the higher that multiple is, the more the activated A_0 amplitude is decreased.

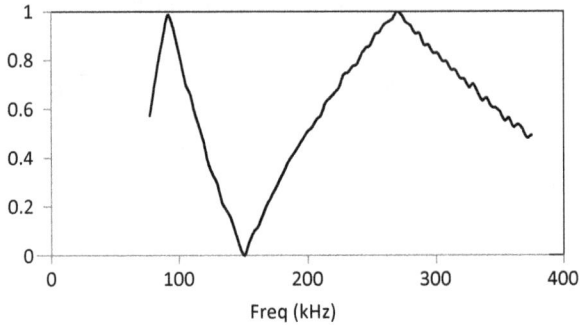

Figure 2.   Lamb wave tuning for the phased array system.

A tuning curve is now obtained in the following manner: based on the wavelength versus frequency curves depicted in Fig. 1, for all frequencies up to 400 kHz, the S_0 wavelength was halved, with these values being subtracted from 3 mm (the selected spacing between consecutive elements in the array) to obtain the optimum PZT dimensions to excite and sense the S_0 wavelengths at their corresponding frequencies. The resulting values were then divided by the A_0 wavelength excited at the same frequencies. Afterwards, the absolute values of the difference between such results and the nearest integer were analyzed and for an easier representation and assessment, these values were doubled and the results were subtracted from unity. Values to be presented close to one represent optimum conditions and values close to zero the worse cases. The resulting tuning curve for the phased array is presented in Fig. 2.

It can be observed that the activation frequencies around 150 kHz (and correspondent wavelengths, array pitch and PZT transducers diameters) should be avoided, while frequencies around 100 and 275 kHz give the best results. It must be remembered, however, that the application of smaller wavelengths is desirable for the detection of smaller damage dimensions, which is achieved at higher frequencies. The frequency of excitation was then selected to be 275 kHz. However, due to the availability of PZT transducer dimensions, PZT transducers with a diameter of 8 mm were used. Considering the spacing between neighboring elements in the array of 3 mm, this corresponds to an array pitch of 11 mm. This should represent half of the wavelength of the S_0 wave to excite. The corresponding actuation frequency, from Fig. 1, is then 245 kHz, with the S_0 waves of that frequency presenting a propagation velocity of 5,450 m/s, in the considered aluminum plate.

PZT transducers (Ceramtec's Sonox P502) with out-of-plane d_{33} were selected to guarantee uniform activation and sensing of waves in the transducer disc plane. Due to its spectrum footprint, reducing harmonic effects and focusing in a single frequency, a Hann windowed sine wave actuation signal was used. In this signal, five sinewave periods are used, with the sinewave having the actuation frequency (245 kHz). The actuation signal has a total duration of 20.41 μs. Since the array elements cannot be used as sensors while actuating, and considering the duration of the actuation signal and the propagation velocity for the generated S_0 waves of 5,450 m/s, a region delimited by a distance of 56 mm around the PZT transducers cannot be inspected.

2.4. *Number of Elements in the Array*

To be able to inspect an entire component, the scans performed by a phased array must be repeated, with the wavefront being steered into different directions in the component in each scan. Consequently, different time delays must be applied between the actuation of consecutive array elements, in different scans. To define the directions and related time delays that must be considered to enable inspection of the entire component, the aperture of the generated wavefronts must be taken into account. The aperture angle is centered around the propagation direction of the wavefront generated by the phased array. It defines the region where such propagating wavefront will be effectively formed and then the useful inspection region for each scan direction.

The aperture ($\Delta\alpha$) is dependent on the total length of the phased array (l) determined by the phased array pitch and number of elements in the array (n) and on the excited Lamb wave wavelength (λ)[23]:

$$\Delta\alpha \simeq 0.886\frac{\lambda}{l} \simeq 0.866\frac{\lambda}{n \cdot \text{pitch}}(\text{rad}). \qquad (9)$$

This relation is valid for all angles of propagation in an isotropic plate and particularly for propagation directions approaching the perpendicular to the linear array. Due to the influence of the array elements in the excitation and initial propagation of the wavefronts, these will present lower amplitudes, decreased inspection capability and will be more scattered in space when their propagation direction approaches the direction of the array, increasing the aperture in those directions. The aperture calculated according to Eq. (9) can be regarded as the minimum aperture. If such value is considered as the angle interval to be established in between consecutive inspection directions, it will be guaranteed that the entire component can be inspected. Simultaneously, the overlap of inspection regions will be promoted and inspection capability will be enhanced.

Theoretically, constructive interference in between generated waves from different transducers in the array is augmented, i.e., the amplitude of generated wavefronts is increased, when the number of generated waves and therefore the number of transducers is increased in the array. However, the generated wavefront will be more scattered in space in those conditions. As can be observed, this results, in fact, in a smaller aperture of the array. As a compromise, it was decided to implement an array with seven elements.

3. Phased Array Experiments

In these experiments, a NI PXI-5421 (single output channel) actuation board and NI PXI-5105 eight channels with simultaneous sampling oscilloscope board (60 MS/s with 12 Bit digitizer) was used for acquisition of the array sensor signals. The two boards were mounted in a NI PXI-1033 chassis connected to a laptop using a PCI-Express card slot.

Besides the use of the oscilloscope board to acquire the phased array sensor signals, this NI system was used to save the test data (sensor data); for data post-processing, including filtering, to implement the damage detection algorithms and to control and synchronize the operation of the entire phased array-based SHM system, particularly in between the developed phased array actuation system and the NI system (regarding the oscilloscope acquisition and the laptop). In these experiments, the NI PXI-5421 single-channel actuation board was only used for synchronization purposes, to send a digital activation signal to the phased actuation system, to the master circuit and to start the actuation of the phased array. This was done when the NI system and particularly the oscilloscope board is prepared to acquire the sensor data. This was enabled by the fact that both NI PXI-5421 and NI PXI-5105 boards are synchronized by the NI PXI-1033 chassis. The oscilloscope is triggered to start saving the test data by the actuation signals sent to the PZT transducers in the array.

As for software, initially NI SIGNAL EXPRESS® was used for a rapid assessment of system performance on signal generation, from each slave board and collectively for the determination of the precision at which time delays are introduced by the master and the subsequent effect on the beamforming process, precise activation of the desirable Lamb waves and wavefronts and sensing and sensor signal acquisition. This software has a visual interface that permits to manage and monitor automatically, in real time and in a simple way, the activation signal to the phased array actuation system and sensor signal acquisition and data recording. It also presents simple but useful data post-processing capabilities for an initial assessment, such as filters, spectrum analyzers, switches, conditional decision making, triggers, mathematical signal operations, etc.

After the previously mentioned initial assessment, the control of the operation of the SHM system, synchronization of the boards, activation of the phased actuation system, sensor signal acquisition, data post-processing and the implementation of damage detection algorithms were performed in a developed dedicated program using LABVIEW®. Virtual Instruments (VI) were graphically coded to control the NI boards; to create the front panel of the instruments and the system, for user interface so that the user can set up and control inspection execution, save data and visualize inspection results. MATLAB® codes were also developed and embedded into the LABVIEW® program to apply data post-processing filters and the damage detection algorithms.

Initial experiments were dedicated to test the slave circuits to verify the generation of the actuation signal waveform, specifically its correct amplitude and frequency, and to assess the introduced errors in time. The complete actuation system, shown in Fig. 3, was then tested to

Figure 3. Phased array actuation system.

verify the correct generation of the wavefront and the influence of the previously referred errors and relative errors in between the different actuation channels.

The outputs from all slave circuits were directly connected to the data acquisition system. The delays introduced by the master circuit were confirmed and the errors in the beamforming process were verified.

In the analysis of all generated signals, the maximum detected difference in amplitude was 610 mV, corresponding to a relative error in amplitude of 5.5% (considering a maximum amplitude of 11 V). The maximum time difference observed between the different highest peaks was 250 ns, corresponding to relative errors in time and in frequency of 2.5% (with relation to an actuation frequency, i.e., inner sine period and frequency, of 245 kHz). Such errors will result in that the wavefront generated will be spread through 1.3 mm in its direction of propagation, considering the selected actuation frequency and the corresponding propagation velocity (of the S_0 waves/wavefronts). It is noted that this wavefront spread is not the same as the wavefront width in the direction perpendicular to its propagation, or aperture. Comparing that value with the excited S_0 wavelength, an error of 6.25% is determined in space.

The linear PZT phased array was tested in a square Al2024-T3 aluminum plate, as illustrated in Fig. 4. The array was applied near and parallel to one of the edges of the plate, in the middle of its width.

The first experiments executed after the implementation of the PZT array were focused on the verification of the correct generation of wavefronts, their propagation directions, amplitudes and apertures. For these initial experiments, a network of three PZT sensors was also used.

Figure 4.   Linear PZT phased array applied to an aluminum plate.

These PZTs were bonded near the other three plate edges (also in the middle of their width), with respect to the phased array. The wavefront was generated and steered to various directions in the plate and, besides the sensor signals from the PZT array, the network sensors signals were also acquired. The time delays introduced by the master circuit were modified in these tests to steer the wavefronts to the network PZT sensors and into their adjacent directions. From these experiments, it was concluded that the wavefronts were successfully generated and steered; their wavelength and frequency were also confirmed.

Through the wavefront ToF, between generation and arrival to each one of the network sensors, the propagating velocity of the S_0 wavefronts was confirmed with the initially calculated values from dispersion curves.

The aperture of the array was then confirmed through the scans in which the generated wavefront was steered to propagate in directions adjacent to the network sensors. For each scan direction, the amplitudes of the resulting acquired signals for each PZT in the network were recorded — the amplitude of the wavefront and the amplitudes of the wavefront preceding and trailing waves. In Fig. 5, the results for the scans performed in directions close to 45° are presented.

For the generation of these scans, a quadrangular, single-layer mesh, with a spacing of 5 mm, was applied to the plate. An array of 300 × 300 (for the 1.5 m × 1.5 m plate) was created, corresponding to each of its positions to one point in the mesh. The amplitude values were then used to populate this matrix. For an easier representation, the amplitudes of the

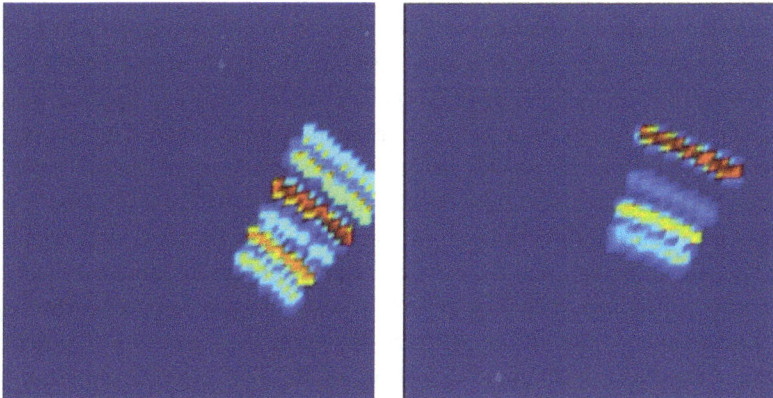

Figure 5.   Verification of wavefront aperture.

wavefront and remaining waves were staggered in space, corresponding to the distances determined through the wave (and wavefront) propagation velocity and the relative time intervals between them in the considered network sensor signal (their arrival time at the sensor). This wave propagation reconstruction process was executed for all scans performed in directions adjacent and including the considered network PZT, and for all three network PZTs.

Finally, the amplitudes of the network sensor signals corresponding to the propagating wavefronts generated by the phased array were assessed. A simple experiment was also performed with all the slave actuation channels deactivated except the one corresponding to the PZT in the array closest to a particular network sensor. This was done to decrease the interference from the remaining PZTs in the array into the generated wave (by the only active PZT), particularly considering its propagation direction connecting the actuator PZT in the array and the network sensor selected. This experiment was repeated for all the three network sensors and the amplitudes of their signals corresponding to the propagating wave were also analyzed, as illustrated in Fig. 6. These amplitudes were then compared to the amplitudes acquired for the wavefront generated by the phased array. It was verified that the sensed amplitudes for the wavefront were ten times higher than the ones obtained for the single-wave generation. As expected, the wavefront amplitude is higher for propagation directions approaching the perpendicular of the array orientation.

3.1. Damage Detection Algorithms

For developing damage detection algorithms, two different approaches were followed concurrently. The first was based on the application of the phased array sensing principle referred in the literature, which is applied by the majority of works in this field. In this approach, since the inspection direction is known, the phased array sensor signals are shifted in time for the corresponding relative time delays imposed in actuation (related to the desired propagation direction). Afterwards, the signals are added, with the resulting signal being analyzed, trying to find a potential damage reflection. With this procedure, incoming reflection waves in the inspection direction will be enhanced, increasing their SNR. After encountering a potential damage reflection — damage detection — its ToF is retrieved from the sensor signals (time between the wavefront generation and detection of such reflection). Knowing the wave propagation velocity (equal to the generated

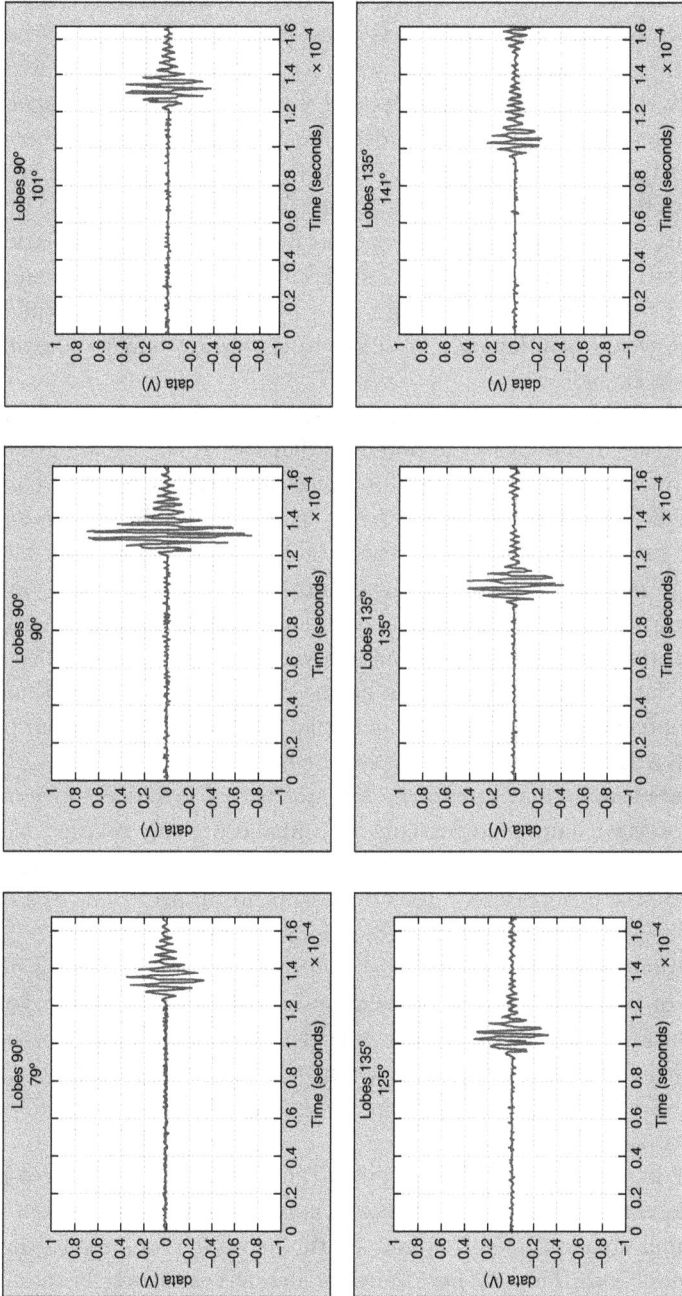

Figure 6.   Wavefront amplitudes detected by the PZT network.

wavefront and reflections propagation velocity), the total distance travelled by the generated and reflected waves is calculated. The distance between the phased array and damage is then determined, as half of the total distance travelled by the generated and reflection waves. Knowing the inspection direction, damage location is then determined. The disadvantage of this approach is that it is based on the approximation that the incoming damage reflection to the phased array is a linear wavefront and not curved, as it is in reality. This approximation gives acceptable results for the determination of damage locations when the damage is sufficiently far from the array and the section of its reflected wave arriving to the phased array is approximately linear. However, when damages are close to the array, damage location is not accurate enough.

This approach can be applied in a supervised or unsupervised method. In the supervised approach, the potential damage reflections are searched for directly in the sensor signals obtained from scans for the current health condition of the component being inspected. There is no need for previously saved signals for a reference state (undamaged or with known damages, to monitor new damages and damage growth) in this mode and then no subtractions or comparisons are done. Since no subtractions are performed, the current signals to be analyzed will present all reflections from boundaries and other intrinsic geometric or material discontinuities in the component. To distinguish a potential damage reflection from these reflections, a supervised approach is required.

In the unsupervised approach, by subtracting reference to current signals, potential damage reflections are enhanced (with relation to the disappearing boundary reflections) and easier to detect in the sensor signals. Besides this, the unsupervised method presents advantages towards a more global automation of the damage detection process.

The application of the supervised approach, depending on the direct detection of potential damage reflections in the current sensor signals, requires such reflections to present considerable amplitudes in order to be identified from the remaining noise in the signal (higher SNR). This is the reason why this method can be applied within a phased array implementation.

Either in the previous method or on the second one, to be presented next, to decrease the random influence of noise in sensor signals, each scan was repeated consecutively 50 times (with the wavefront propagating in the same direction). This was performed for all scans executed (for the entire component, considering all inspection directions), both for the reference and actual health conditions, except for all corresponding sensor signals. After performing the multiple repeated scans, sensor signals corresponding to each transducer and for each scan direction are averaged, and maximum and minimum values are retrieved for all times. Based on a Gaussian distribution approach, new maximums and minimums corresponding to a 90% probability that the signal will be contained in such range are determined, for all times. Signal bands are then obtained for all sensor signals corresponding to the actual and reference conditions of the component, for all scans, i.e., inspection directions. Specifically, for the unsupervised approach, the comparison of sensor signals for current and reference states, for each of the different scans (and for each sensor), is then based on the verification of when the current health state corresponding signal band differs from the reference band by more than 50% of its width.

In this damage detection procedure, all possible reflections are considered. To reduce further the influence of (random) noise and resulting false indications, possible damage locations determined for each reflection, for each sensor signal and for each inspection scan are superimposed. This is performed considering the sensor data for all activations of the array for the same and different directions of inspection. This is equivalent to performing the intersection of all possible damage locations and forming a damage index, with the indication of damage positions being enhanced with respect to false-positives, since the probability of intersection of random false indications resulting from (random) noise to fall in the same position is reduced. In this process, bandpass filters are also applied to sensor signals, around the elected frequency for the application of the method (actuation and excited waves and reflections frequency), as before.

The second procedure was derived from the ToF determination of damage reflections, considering the excited wavefront and the wave reflection generated by the damage (in this case considered as a curved circular wave propagating and not as a linear wavefront). The procedure is depicted in Fig. 7.

For each transducer in the array and for each inspection direction — considering the multiple scans performed repeatedly and consecutively for each direction — its sensed signal band is searched for potential damage reflections. In this procedure, the ToF determined concurrently by the first method is also used. As referred previously, the ToF of all reflections are considered, in this case until the ToF of the reflection of the farthest boundary (corner) with relation to the phased array is reached since the damage is contained within the component.

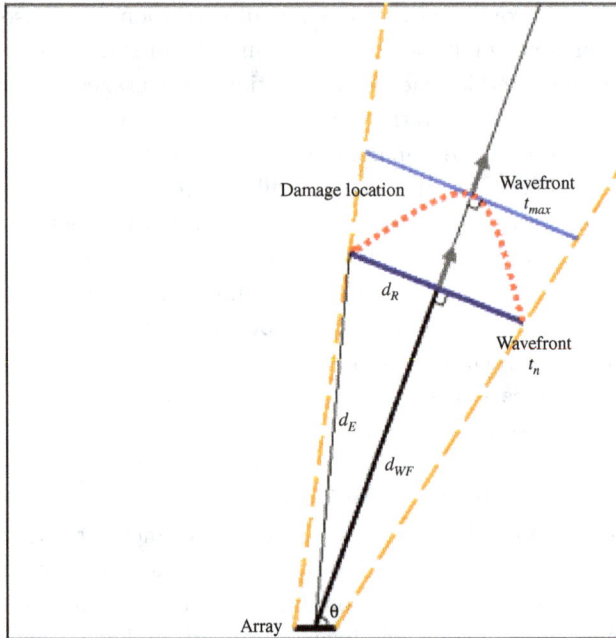

Figure 7.   Developed damage location algorithm for the phased array SHM system.

Since the signals from the individual sensors are considered in this method, although the phased actuation is applied, the reflections to be searched for in the sensor signals will not be directly enhanced by the phased array sensing method. A baseline or unsupervised approach was adopted, since it will likely give better results with such non-enhanced reflection amplitudes. Formed sensor signal bands from the multiple consecutive actuations for a single inspection direction are therefore subtracted from baseline signal bands or signal bands corresponding to reference health state (obtained and saved from previous scans). Outliers are then determined and the ToF for all outliers is established.

In this damage detection technique, the known direction of propagation of the wavefront (predetermined by the phase delays) is marked for each sensor in the array for the specific scan. All the ToFs considered are then related to the corresponding total travelled distance of the activated wavefront plus reflected wave, through the known wave propagation velocity. As mentioned before, a damage distance, corresponding to half of the total traveled distance, is determined and marked in the inspection direction lines for all PZTs, corresponding to the intersection point of the line representing the wavefront propagation direction and the line marked as wavefront at t_{max} (in light blue), for the central PZT. Departing from each of those initial points and through the ToF, parabolas are determined for each PZT. Each parabola is symmetric with relation to the inspection direction line passing through the corresponding transducer. It has its focus point in the corresponding transducer and its directrix parallel to the generated wavefront at a distance from the transducer corresponding to the ToF previously determined considering the known wavefront propagation velocity. The intersection of all the parabolas will determine or enhance possible damage locations, eliminating or decreasing the relative importance of false indications.

A mesh and corresponding array was subsequently generated, similar to what was described in the previous section. The positions in the array corresponding to points in the mesh that form previously described parabolas are filled cumulatively with a unit value, for each parabola they are part of. To decrease the processing time, only the points in the mesh within the inspection region, delimited by the aperture of the wavefront and by the wavefront at t_{max}, are considered. After all inspection directions being scanned, the different arrays are added, considering that the different directions of inspection were selected to promote a slight superposition of the different corresponding inspection regions. This enables the determination for each point in the mesh of a measure of the probability of damage existence, relative to the other points in the mesh.

4. Damage Detection Experiments

The developed phased array system was experimentally tested in the 1.5 m × 1.5 m × 2 mm aluminum plate, subjected to different boundary conditions, ranging from having the plate being totally supported on one of its faces (to assess the system for higher Lamb wave propagation damping), to simply supported and riveted in the boundaries. The experimental setup is presented in Fig. 8. Experiments were conducted first in a laboratory setting and afterwards in an aircraft maintenance hangar, with no surrounding noise control and limited temperature control. Experiments were performed with the cumulative introduction of surface and through the thickness circular holes and cuts with different orientations, with a maximum dimension of 1 mm to simulate damage. Damages were not

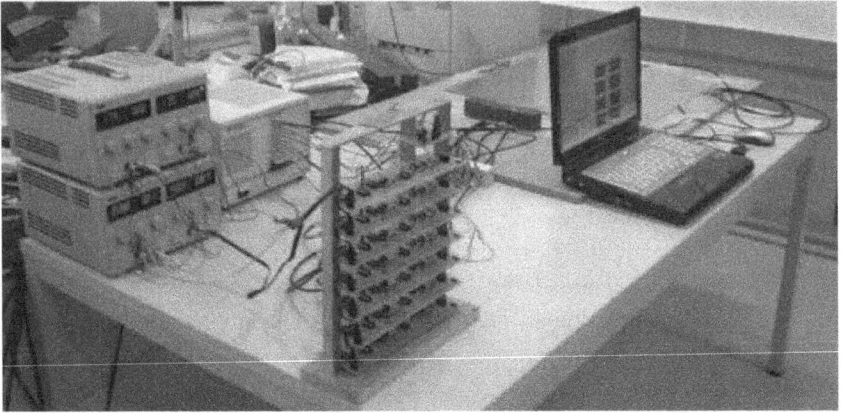

Figure 8.   Phased array experimental setup.

inflicted around the phased array, respecting the non-inspected distance of 56 mm.

A total of 46 damages were introduced cumulatively in different positions in the entire aluminum plate. From these, 17 damages were simulated by through the thickness holes, seven damages by surface holes, nine damages by surface cuts — three oriented in a radial direction considering the phased array as the center; two oriented perpendicular to the local radial direction, one at 45° from the local radial direction and three randomly oriented, at an angle between 30° and 60° with the local radial direction. Thirteen damages were simulated by through the thickness cuts — five oriented in a radial direction, two oriented perpendicular to the local radial direction, two at 45° from the local radial direction and four randomly oriented, at an angle with the local radial direction.

The imposed damages were successfully detected, with the exception of damages created behind other damages with respect to the phased array (in the same radial direction). Also, the relative probability of detection was lower for 40% of the cuts aligned with wavefront propagation directions, since their thickness was lower than 1 mm.

Software for automated inspection was developed in LABVIEW®, with embedded MATLAB® codes for signal processing and to implement damage detection algorithms.

To enable the user to verify the time delays introduced by the generation system in the phased actuation and the corresponding inspection direction, a window was also created with that objective.

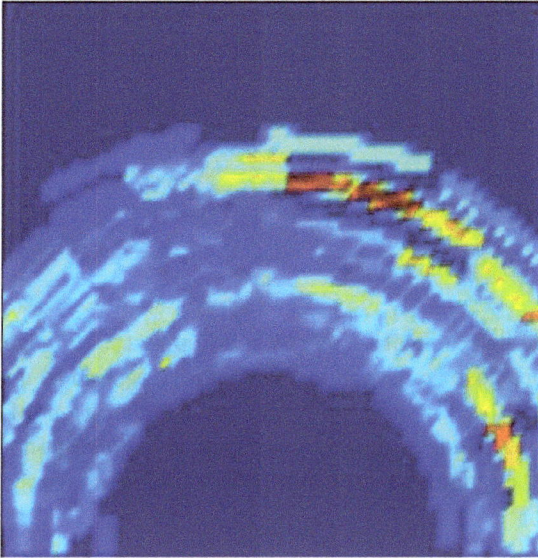

Figure 9.   Damage location contour output.

A window was then created for the user to set and trigger the inspection system. In this window, the user can define the mesh spacing and possible post-processing algorithms to be applied. The folders to save data in the different steps of the data post-processing procedure are established by the user, including the folder where the final results will be saved.

A typical results window from the software is shown in Fig. 9. In this plot the superimposed solution (tomography based) is presented to the user — as the plot of the final matrix obtained through the method explained at the end of the previous section. In this particular case, the presented results were obtained for an enlargement of an existing through the thickness circular hole, from 0.5 to 1 mm in diameter. The system pinpointed the location in the position presented as dark red in the plot. This result is presented to prove the capability of the system to monitor damage growth (in steps lower than 1 mm).

In Table 1, the six locations with highest probability of damage existence are also presented. The relative probability non-dimensionalized to unity is presented. Here, the coordinate system origin is positioned in the center of the phased array, in the middle of the lower boundary, with the X-axis being parallel to the lower edge of the plate, pointing to the right, and the Y-axis being perpendicular to the array, pointing upwards.

Table 1.   Probability of detection.

X(m)	Y(m)	Probability
0.26	1.0	1.00
0.22	1.1	0.82
0.63	0.3	0.60
0.72	0.7	0.55
0.55	0.9	0.50
0.50	0.8	0.45

Besides the first indication, corresponding to the damage location, the remaining indications correspond to false-positives, due to noise, boundary reflections and the intrinsic and inevitable generation of other waves with smaller amplitude than the wavefront and corresponding generated reflections in the damage location, where mode conversion could also happen.

5. Concluding Remarks

A SHM system based on a dedicated phased array actuation system, required for the generation of the waves and wavefront, has been developed. This system is based on a master MCU circuit, controlling the phased activation of slave channels. The master MCU has programmed time delays to be applied for the phased activation for all scan directions to be implemented. Furthermore, the number of scans and scan repetitions have also been determined. The slave circuits/channels, connected to each PZT in the array, generate the correct actuation signal waveform. In these circuits, an output switch was applied to match the impedance of actuation and acquisition systems, which is a usual problem in the existing systems. With the use of such switch, measured wave amplitudes were increased significantly.

The final actuation system was tested. The correct determined inspection directions and the array aperture were verified. The actuation system presented errors lower than 6% and 3%, respectively in the relative amplitude and time (and frequency) of excited waves and wavefront generation.

Experiments were performed to validate the performance of the phased array SHM system. An aluminum plate 1.5 m × 1.5 m × 2 mm was used, as representative of most aircraft structural components since these can be divided in the majority of cases into plate-like or low curvature-like shells. The plate was subject to different boundary conditions ranging from totally supported (to increase Lamb wave propagation damping) to simply supported and riveted in the boundaries.

For the experimental setup, the NI PXI-5105 data acquisition system with eight channels with simultaneous sampling oscilloscope was used, and LABVIEW® code (with embedded MATLAB® codes) was developed to implement the interface for the SHM system and the damage detection process. This code performs signal processing and applies the developed damage detection algorithms in an automatic manner.

Experiments were performed with the introduction (cumulatively) of surface and through the thickness circular holes and cuts (with different orientations), with a maximum dimension of 1 mm, to simulate damage. The simulated damages were successfully detected. The only exceptions were damages created behind other damages previously introduced, i.e., in the same radial direction, with relation to the phased array (originating in the phased array). It is noted that cuts aligned with wavefront propagation directions were more difficult to detect, since their thickness was inferior to 1 mm.

References

1. Wurstenberg, H., B. Rotter, H. Klanke, and D. Harbecke, Ultrasonic phased arrays for non-destructive inspection. *Materials Evaluation* **51**(7) (1993) 669–671.
2. Bao, J., Lamb wave generation and detection with piezoelectric wafer active sensors, Ph.D. Thesis, College of Engineering and Information Technology, University of South Carolina, 2003.
3. Purekar, A. S., Piezoelectric phased array acousto-ultrasonic interrogation of damage in thin plates, Ph.D. Thesis, University of Maryland, College Park, 2006.
4. Pena, J., C. Perez, R. Martinez-Ona, Y. Gomez-Ullate, F. M. Espinosa, and G. Kawiecki, Phased array transducers for damage detection in aircraft structures. *Fatigue and Fracture of Engineering Materials and Structures* **31**(2) (2008), 1019–1030.
5. Malinowski, P., T. Wandowski, I. Trendafilova, and W. Ostachowicz, A phased array-based method for damage detection and localization in thin plates. *Structural Health Monitoring* **31** (2009) 611–628.
6. Salas, K. I. and C. E. S. Cesnik, Guided wave experimentation using CLoVER transducers for structural health monitoring. In *49th AIAA/ASME/ASCE/AHS/ASC Structures, Structural Dynamics, and Materials*, Schaumburg-IL, USA, 2008.

7. Michaels, J. E. and T. E. Michaels, Guided wave signal processing and image fusion for *in situ* damage localization in plates. *Wave Motion* **44** (2007) 482–492.
8. Sedlak, P., Y. Hirose, S. A. Khan, M. Enoki, and J. Sikula, New automatic localization technique of acoustic emission signals in thin metal plates. *Ultrasonics* **49** (2009) 254–262.
9. Sohn, H., H. W. Park, K. H. Law, and C. R. Farrar, Combination of a time reversal process and a consecutive outlier analysis for baseline-free damage diagnosis. *Journal of Intelligent Material Systems and Structures* **18** (2007) 335–346.
10. Xu, B. and V. Giurgiutiu, Single mode tuning effects on Lamb wave time reversal with piezoelectric wafer active sensors for structural health monitoring. *Journal of Non-Destructive Evaluation* **26** (2007) 123–134.
11. Gomez-Ullate, Y., F. Espinosa, P. Reynolds, and J. Mould, Selective excitation of Lamb wave modes in thin aluminium plates using bonded piezoceramics: FEM modelling and measurements, Poster 205, ECNDT, 2006.
12. Gao, W., C. Glorieux, and J. Thoen, Laser ultrasonic study of Lamb waves: Determination of the thickness and velocities on a thin plate. *International Journal of Engineering Science* **41** (2003) 219–228.
13. Paget, C. A., S. Grondel, K. Levin, and C. Delebarre, Damage assessment in composites by Lamb waves and wavelet coefficients. *Smart Materials and Structures* **12** (2003) 393–402.
14. Yuan, S., D. Liang, L. Shi, X. Zhao, J. Wu, G. Li, and L. Qiu, Recent progress on distributed structural health monitoring research at NUAA. *Journal of Intelligent Material Systems and Structures* **19** (2008) 373–386.
15. Rizzo, P. and F. L. di Scalea, Feature extraction for defect detection in strands by guided ultrasonic waves. *Structural Health Monitoring: An International Journal* **5**(3) (2006) 297–308.
16. Monnier, T., Lamb wave-based impact damage monitoring of a stiffened aircraft panel using piezoelectric transducers. *Journal of Intelligent Material Systems and Structures* **17** (2006) 411–421.
17. Chang, F. K., F. C. Markmiller, J. B. Ihn, and K. Y. Cheng, A potential link from damage diagnostics to health prognostics of composites through built-in sensors. *Journal of Vibration and Acoustics* **129** (2007) 718–729.
18. Garg, A. K., D. R. Mahapatra, S. Suresh, S. Gopalakrishnan, and S. N. Omkar, Estimation of composite damage model parameters using spectral finite element and neural network. *Composites Science and Technology* **64** (2004) 2477–2493.
19. Ihn, J. B. and F. K. Chang, Pitch-catch active sensing methods in structural health monitoring for aircraft structures. *Structural Health Monitoring: An International Journal* **7**(1) (2008) 5–19.
20. Raghavan, A. and C. E. S. Cesnik, Effects of elevated temperature on guided-wave structural health monitoring. *Journal of Intelligent Material Systems and Structures* **19** (2008) 1383–1398.

21. Wang, L. and F. G. Yuan, Damage identification in a composite plate using pre-stack reverse-time migration technique. *Journal of Intelligent Material Systems and Structures* **12** (2001) 469–482.

22. Yu, L., G. Santoni-Bottai, B. Xu, W. Liu, and V. Giurgiutiu, Piezoelectric wafer active sensors for *in situ* ultrasonic-guided wave SHM. *Fatigue & Fracture of Engineering Materials & Structures* **31**(8) (2008) 611–628.

23. Giurgiutiu, V., *Structural Health Monitoring with Piezoelectric Wafer Active Sensors*. Boston, USA: Elsevier Academic Press, 2008.

Chapter 6

SHM of Composite Structures by Fiber Optic Sensors

Alfredo Guemes

Department of Aeronautics, Universidad Politecnica Madrid, SPAIN
alfredo.guemes@upm.es

Fiber optic sensors for strain measurements, particularly fiber Bragg grating (FBG) sensors, have been used for the last 20 years, building up confidence in their performance. FBGs can measure strain with accuracy similar to the standard strain gages and extensometers and they are also comparable in many aspects from a user's point of view. Indeed, their measurements are local and directional, they require compensation for temperature and they are commonly bonded onto the surface, but also it is possible to put the optical fiber embedded into the laminate in the case of composite structures. Nevertheless, structural health monitoring (SHM) means mainly damage detection. Typical damages, like a local crack, act as failure initiation points, which may reduce the strength of the structure by 50%, but they do not change the global strain field. Only a small region around the crack is affected. This chapter reviews some procedures to get information about damage from strain measurements, an issue that is far from being solved, but some promising approaches have been proposed and demonstrated.

1. Introduction

The state of the art of the three main components for this chapter (fiber optic sensors, composites and SHM) is reviewed in this section. Only a quick review is done to highlight the relevant points. Readers are referred to the widely available literature for more detailed explanations.

1.1. *Fiber Optic Sensors (FOS)*

The fiber optics technology started in the 1970s for long distance communications and 10 years later, the first applications of FOS started to appear in different fields; chemical sensors is probably the field with a wider set of applications, but strain–temperature measurements have grown significantly during the last decade.

FOS are classified by the sensing principle: intensity, phase or interferometer sensors, polarization and wavelength. Those based on wavelength (fiber Bragg grating, FBG) are currently preferred, because they offer multiple advantages: easy multiplexing capability, long-term durability and immunity to power interruptions. Two main events may be identified as milestones to understand the progress of the technology: (i) initiation of mass production of FBG on photosensitive fibers by phase mask procedures in 1989[1] and (ii) publication of a procedure for a very high multiplexing capability, more than 1,000 gratings per fiber[2] based on optical frequency domain reflectometry (OFDR) in 1999. Since then, the two main streams for research were opened.

(a) FBG as point sensors (typical 10 mm long), engraved at the core of an optical fiber; easy multiplexing in wavelength, up to 10–15 sensors/fiber, depending on the wavelength window of the interrogation system.
(b) Distributed sensing, a unique possibility for getting a continuous reading of strain/temperature all along the optical fiber, with a spatial resolution of a few mm. They work with plain optical fibers, but in some cases, sensitivity and acquisition rate may be improved with special fibers.

Most of the existing textbooks on SHM,[3,4] just as examples of recently published books, devote a chapter for describing how FOS work, which is the starting point, but it needs to be pointed out that FOS are simply strain–temperature sensors, and something else is needed to get information about damage, which is the goal of SHM.

1.2. *Structural Health Monitoring (SHM)*

SHM is defined as the process of acquiring and analyzing data from on-board sensors to evaluate the health of a structure. It shares the objective of detecting local damages, but while non-destructive testing (NDT) techniques need external instruments, like X-ray sources and detectors, or ultrasonic probes, which imply that the structural element needs to be accessed and probably dismounted for inspection, for SHM, the sensors responsible for getting information from the structure are permanently attached, at fixed positions. The structure can be remotely inspected, without disassembly, without specialized personnel and with significant savings in maintenance costs. Sensors have to be distributed through the structure, because the location of damage occurrence is usually unknown. Consequently, the sensors need to be small in size and cheap to allow for tens of hundreds of sensors for each structural element. PZT and FOS sensors are the best candidates and the more widely studied, but alternative solutions like MEMS are also feasible.

The raw information acquired by the sensors needs to be processed to distinguish the perturbations caused by damages from environmental disturbances on the signals, leading to the four levels of SHM: identification of damage occurrence, localization, classification of damage and quantification of damage size.

Prognostic is frequently added to the diagnostic, as the fifth level of SHM. It includes the statistical forecast of the remaining life, based on the real condition of the structure and the actual loads. Here again, the strain readings acquired by the FOS are required to feed into these models.

1.3. *Composite Structures*

There is no need to highlight the advantages and limitations of advanced composite materials for low-weight structures: higher specific properties, particularly higher stiffness, tailored designs, fatigue and environmental resistance. As disadvantages, its lower damage tolerance, limited reparability and frequently higher costs are mentioned.

Dealing specifically with the usage of FOS for composite structures, two approaches are feasible: sensors bonded to the cured structure, or sensors embedded into the composite material during the manufacturing phase.

The first option, externally bonded sensors, does not pose any special issue other than protecting adequately the sensors and the optical fiber for long-term endurance. A careful analysis of packaging and mounting FBG was done by Grabovac *et al.*[5] and it is available as a public report from Defense Science and Technology Organization (DSTO), Australia.

It has always been sought as the smartest solution to build the structure with embedded sensors, which brings obvious advantages for long-term durability and aesthetic appearance, but it is more difficult to implement. This is a list of the difficulties to address:

(1) The optical fiber may break during the curing stage, because it is rather brittle, particularly when curing is done in an autoclave. A careful

design is needed to avoid sharp steps that the fiber is forced to follow (i.e., if the optical fiber is located at a 10th ply, and nothing is done at the edge to protect the exit, the optical fiber would break at the sudden step of $10 \times 0.15 = 1.5$ mm). Tooling displacements, even the small ones caused by thermal expansion, are also a frequent cause for optical fiber failure.

(2) When the optical fiber is embedded between fabrics, the microbending caused by the fabric undulation will fade out the optical power after a few meters. This effect is easily avoided by adding a narrow prepreg tape accompanying the optical fibers to smooth the path.

(3) The diameter of the polyimide-coated optical fiber (150 microns) is much higher than the diameter of graphite fibers (5 microns) and roughly similar to an individual prepreg ply thickness. There are no problems of losses in the mechanical properties of the host material as far as the OF is aligned with the ply direction (Fig. 1); when the OF crosses orthogonally the reinforcing plies, resin "eyes" appears at both sides, which may initiate fatigue cracks.

(4) Once the material is cured, the internal residual stresses caused by the anisotropic shrinkage may cause a distortion of the spectral response of the FBG (Fig. 2). Two kinds of problem may occur:

    (a) Peak splitting caused by the transversal stresses on the FBG. On the free sensor, the two polarization modes of the light travelling through the core were superimposed (optical properties at the two transverse y and z directions were identical), but with the existence of in-plane stresses in the y direction, which slightly

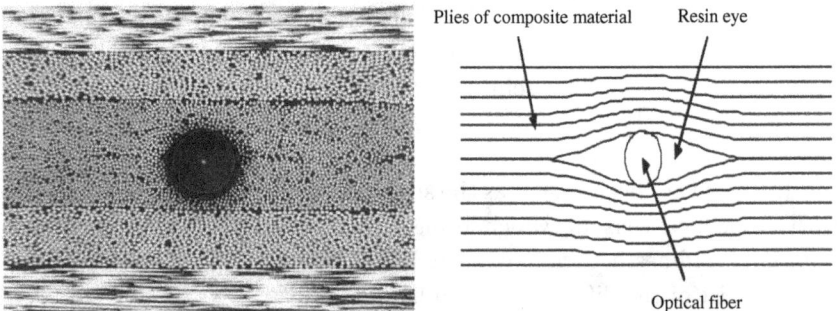

Figure 1. Micrography of an optical fiber embedded parallel to the graphite fibers of a tape laminate. On the right side, sketch of a laminate, showing the "resin eye".

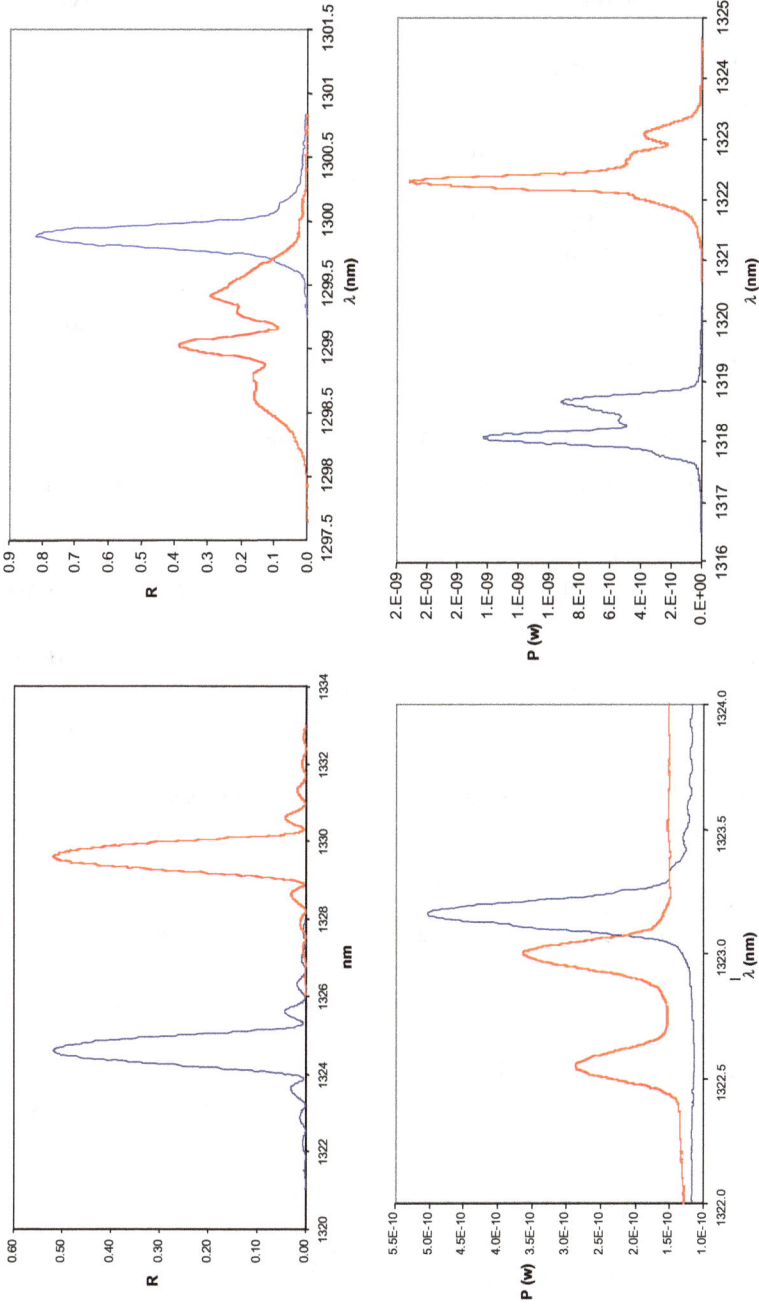

Figure 2.   Upper left: Free FBG submitted to uniaxial strain. Upper right: Longitudinal strain gradients along the grating. Lower left: Peak splitting caused by transversal stresses on the grating. Lower right: Distorted spectrum commonly observed for embedded FBGs.

changes the refractive index (n) in this direction, it will also change the condition of peak reflected wavelength $\lambda = 2n\Lambda$.

(b) Peak distortion caused by the strain gradients along the sensor length (x direction).

(5) Ingress/egress of the OF from the laminate. The strength of the optical fiber itself is low, less than 50 N, so that it may be easily broken during the demoulding and trimming operations of the cured parts. Several "laboratory solutions" are available, like protecting the fiber with a Teflon mini-tube. It is also embedded several centimeters inside the laminate, but the most robust industrial solution is surface-mounted microconnectors.

2. Damage Detection with Fiber Optic Sensors

In spite of the large number of papers related to the topic of damage detection by FOS, only a few address the real issue that measuring strains is not the same as detecting damage. It is important to recognize that damage is not a physical parameter, but just a local change in the material's properties or at the structure boundaries (a crack is simply a new boundary), which degrades the structural performances. Damage can only be detected by comparing the response of the structure to some external stimulate (forces, elastic waves, thermal heating and so on), before and after damage occurrence.

As such, there are no sensors that may directly detect damage; there is always a need for algorithms to translate the raw information obtained by the sensors. A local damage may be the failure initiation point, may decrease in a large percentage the strength of the structure, but before the catastrophic failure, it produces negligible changes in most of the parameters of the structure (natural frequencies, global strain fields and so on). Most of the work done for SHM is just to identify the "features", or parameters that are sensitive to minor damages and that can distinguish the response from natural and environmental disturbances.

As regards FOS, the first ideas for damage detection appeared as early as 1985 by checking the continuity of the optical fiber.[6] This approach was not robust and sensitive enough, and it was not pursued. From 1996, a series of articles from Takeda dealt with detecting damage in composite materials by analyzing the spectral distortion of the FBG peak caused by the damage. The basic argument is that a local damage, like a delamination, will produce local strains that will be detected by an FBG, if located just on the damage. From spectral distortion, the strain gradients may be predicted.

The basic limitation for this approach is that damage must happen just besides the FBG; only when the damage location is known *a priori*, the method has a practical usefulness. A case for application was the detection of the stiffeners "run away" in composite stiffened panels (debonding starting at the tip of the stiffeners), explained in Section 5. For the general case, when damage may happen somewhere on the structure, three approaches are feasible, as illustrated in Fig. 3.

2.1. *Pattern Recognition Applied to Strain Measurements*

Pattern recognition techniques are numerical methods for handling large data sets for finding similarities and hidden relationships. When applied to the strains acquired on the pristine and on the damaged structure, they may reveal the little strain changes associated with a damage. The detection capability will be related to the number of strain sensors, which implicitly establish the distance to the crack to the nearest sensors, and it is also related to the damage size. This approach, with examples on real structures, as a wind turbine blade and a satellite structure, is detailed in Section 3.

2.2. *Distributed Sensing*

This technique obtains the strain along the whole length of the optical fiber. Of course, the former algorithms may also be applied for these strains, but a simpler approach is proposed here. When damage occurs at the path of the optical fiber, the damage will be detected because a local damage, even a delamination, always causes a local change in the strain around the damage by relaxation of internal strains. An important difference with the former approach is that the structure does not need to be externally loaded. It is only the relaxation of stresses that is detected. Strains will remain at zero in most of the fiber, except at the damage area.

Figure 4 plots the strain field, acquired by a single optical fiber covering the whole surface of a flat laminate after it was damaged. The clear strain peak at the location of damage can be seen. The plate was impacted to cause delamination. By covering the whole surface of the plate, it enables detecting damage anywhere in the structure. However, this does not seem to be a realistic approach for real structures. The idea of "distributed

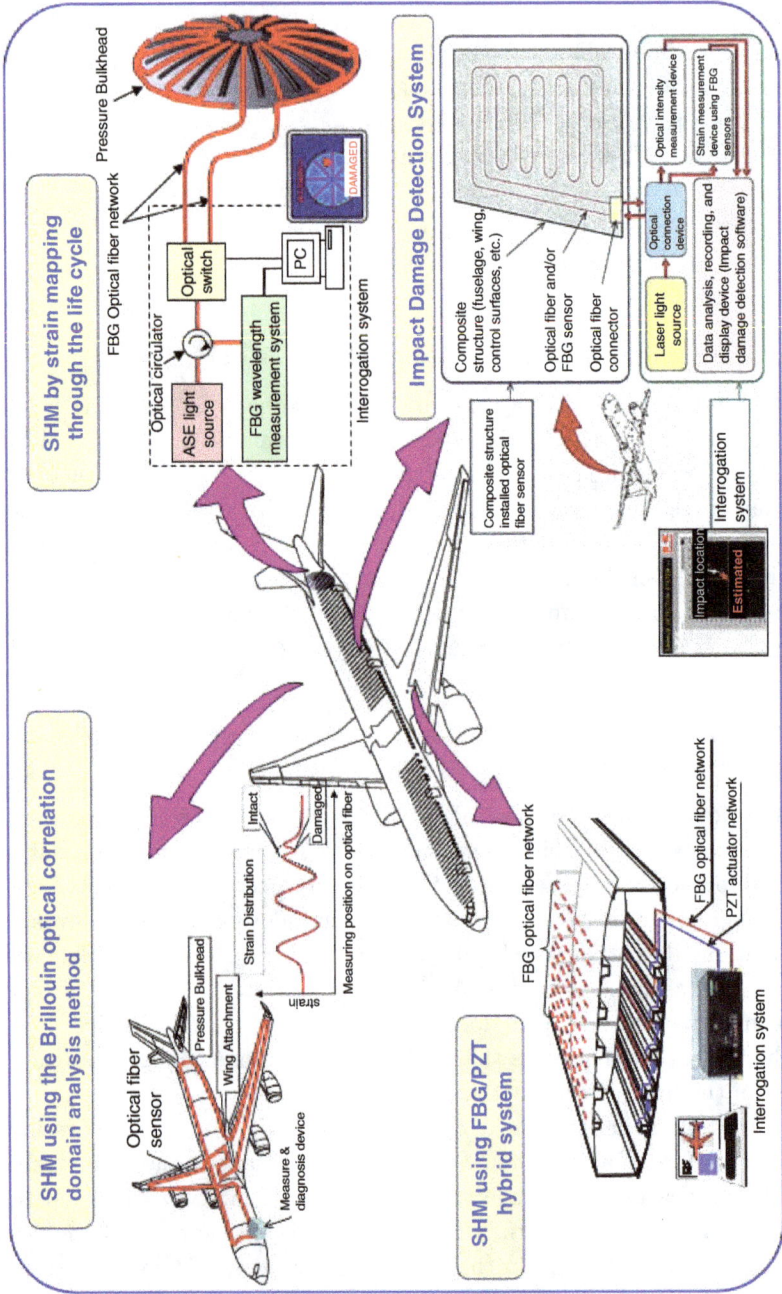

Figure 3. FOS-based SHM systems in the METI Project (from Ref. [7]).

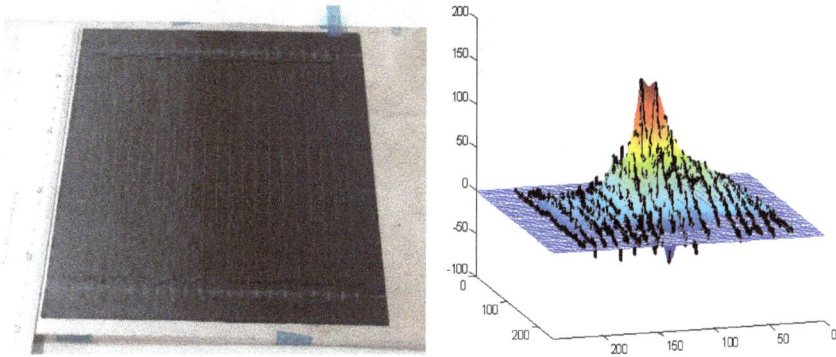

Figure 4.   Laminate with a single optical fiber covering the whole surface, and composed image from the obtained residual strains caused by impact damage.

sensing" is particularly useful to detect damage near the laminate edges, like "man holes" and door surrounding areas, which are areas more prone to damage by impacts. Examples with experimental results are given in Section 4.

2.3. *Detecting Guided Waves by FBGs*

Conventional FBG interrogation systems work at low frequencies; 2 kHz is a typical value. This value is strongly dependent on the kind of interrogation scheme see Ostachowicz and Güemes's book for a wider discussion on available alternatives. Also, the sensitivity of the commercial equipment is about 1 microstrain, similar to electrical extensometer; both parameters are enough for most practical cases, including mechanical vibrations of structures, but cannot acquire the transient signals associated to elastic waves, coming from an impact or by acoustic emission sources. These elastic waves are characterized by higher frequencies (typically up to 500 kHz), and weaker strain level (1 picostrain). The limitation is not due to the FBG sensor itself, able to follow elastic waves up to 1 MHz, with a maximum sensitivity when the fiber is aligned with the wave propagation direction, and zero when the wave impinges orthogonally to the sensor. Several authors have developed laboratory equipment for acquiring high-frequency signals, most of them based on a ratiometric approach.[8] The acquired signals are comparable to what is achieved with PZT wafers, which has pushed the idea of hybrid PZT–FOS active damage detection systems (Fig. 5), similar to the conventional PZT active networks, but FBGs are

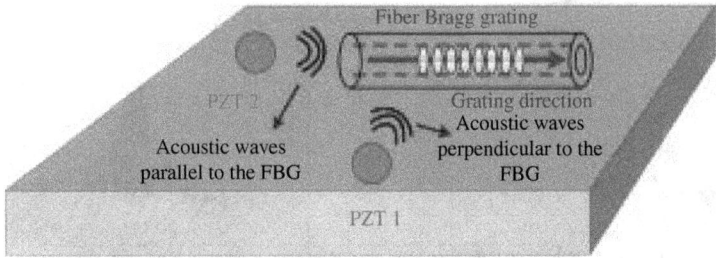

Figure 5.   Hybrid FBG-PZT system for SHM (from Ref. [8]).

employed as sensors to leverage their lower mass and size, directional sensitivity and multiplexing capability.

A main limitation of the ratiometric approach is the need to keep the sensing FBG tuned with the interrogation window, for which there must be an overlap of the spectral peaks; a drift caused by temperature or static strains will bring the sensors out of the operating window, which is a strong practical limitation. To overcome this issue, Krishnaswamy et al.[9] proposed to use a photorefractive crystal (PRC) to split the low- and high-frequency parts of the signals. Several companies are now commercializing products based on this approach, which seems to be the way for ultrasonic detection with fiber optic sensors.

3. Strain Field Pattern Recognition Techniques

Strain collections obtained after several experimental measurements allow the creation of data sets that can be seen as patterns. The study of these groups leads to damage detection based on pattern recognition techniques. The features extraction can be defined as the process of identifying damage-sensitive parameters from the gathered data. This process usually results in some form of data reduction.

The patterns can be continuous variables, discrete variables or a combination of both and, can be expressed in the form of vectors, matrices or multi-dimensional arrays. When pattern recognition techniques such as damage detection approach are used, it must be assumed that each pattern represents a particular structural state. The main idea is then to determine whether a structure is damaged or not and try to assess the damage severity.

For example, when strain field pattern recognition techniques are used, the approach is to correlate all the strain measurements gathered from a network of sensors and to discern if something has changed, in particular, the global stiffness and the strain field between different sensors, as a consequence of damage occurrence in the structure.

There are two classical categories for damage detection by means of pattern recognition. The first approach includes the so-called statistical methods and the second approach includes the so-called syntactic methods. Statistical methods assign features to different classes using statistical density functions, whereas the syntactic methods classify data according to its structural description. Statistical modeling requires a previous statistical characterization of data, before any statistical inference can be reached. The statistical methods are most widely used in SHM. Many techniques for statistical analysis have been developed for building models under uncertain conditions. However, in SHM applications, all the measurements must be studied together in order to increase the probability of damage detection. Then, it is necessary to use multivariate statistical tools in order to obtain some valuable information about the system behavior.

Usually, multivariate data are grouped in batches. Data processing in batch or semi-batch consists of measuring different variables as a function of time.

In order to perform online monitoring of multivariate data, diagnostic and fault detection, several methods have been reported in the literature. These methods are known as multivariate statistical projection methods. Data are projected onto a lower-dimensional space using specially designed mapping functions. This process is called data reduction or data condensation. One of the most used projection methods is the principal component analysis (PCA). PCA provides arguments on how to reduce complex data set to a smaller dimension and also reveals simpler patterns or "structures" that may be hidden under the data. The ultimate goal of the technique is to discern which data represent the most important dynamics of a particular system and which data are redundant or just noise.

PCA is a mathematical procedure that uses an orthogonal transformation to convert a set of observations of possibly correlated variables into a set of values of linearly uncorrelated variables called principal components. This transformation is defined in such a way that the first principal component has the largest possible variance (i.e., accounts for as much of the variability in the data as possible). Usually, the number of principal components can be much smaller than the number of original variables. Each succeeding component, in turn, has the highest variance possible under the constraint that it be orthogonal to (i.e., uncorrelated with) the preceding components. The PCA algorithm is composed of the following mathematics operations (MATLAB® tools are available for it):

(1) Organize the data set as $n \times m$ matrix, where n is the number of experiments and m is the number of measured variables: X.
(2) Normalize the data to have a zero mean and unity variance.
(3) Calculate the eigenvectors and eigenvalues of the covariance matrix: $C = XX^T$.
(4) Keep only the first eigenvectors as the principal components.
(5) Project any new collected data into the former baseline.
(6) Identify if new data follow global trends (damage index).

There are statistical tools that, used along with PCA, allow the detection of anomalous behavior in systems. The two most common tools are the Q index (or index squared prediction error, SPE) and the T-index. The index Q indicates how well each sample fits the PCA model. It is a measure of the difference between a sample and its projection in the main components retained by the PCA model.

To demonstrate the power and limitations of the technique, two real cases have been analyzed:

3.1. Composite Wind Turbine Blade, 13.5 m Long

The composite wind turbine blade is instrumented with four optical fibers, six sensors at each fiber, with a distance of 2 m between the sensors (Fig. 6).

The blade was statically loaded, as shown at Fig. 6; first, as it was manufactured (baseline response) and later after producing damages of increasing severity, as a debonding of 100 mm at the trailing edge (D1), which minimally alters the general strain field acquired by the sensors, to a transverse cut 50 mm long D3, which produces a much bigger change. Figure 7 shows the damage index calculated by PCA from the experimentally obtained strain measurements (each experiment means a data acquisition with a different load).

It is clearly seen how the damage index allows the detection without uncertainty not only the large damage case D3, but also the trailing edge debonding D1. Nevertheless, it is important to realize that under realistic conditions (strain measurements taken every 2 m, as distance of sensors),

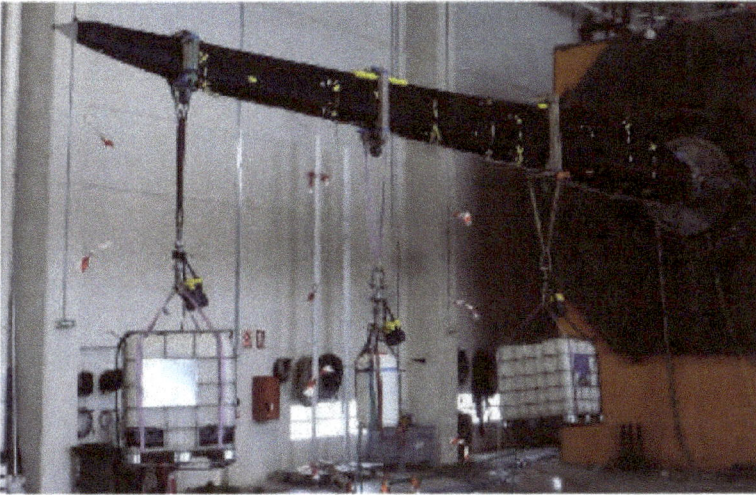

Figure 6.   Composite wind turbine blade instrumented with 4 × 6 FBG sensors.

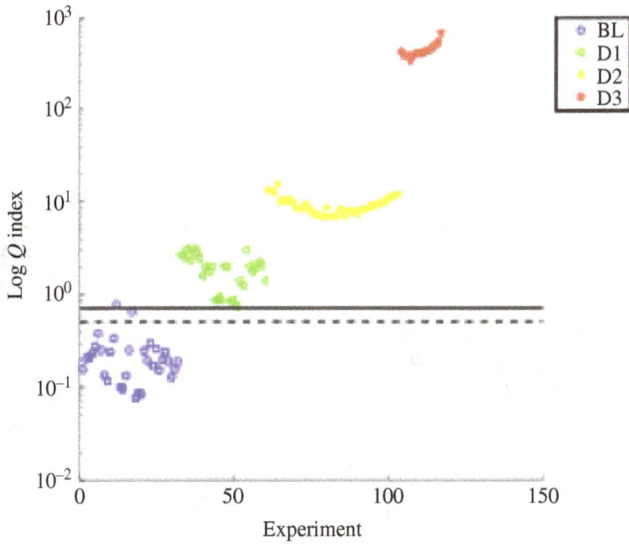

Figure 7.   Damage index at the wind turbine blade after damages of increasing severity.

the debonding needs to be relatively large (100 mm) to be detected without ambiguity.

3.2. *Second Experiment: CFRP Isogrid Structure*

- An open isogrid structure, also known as lattice structure, made of CFRP is a weight-efficient structural concept.
- One element was built by fiber placement at the "CASA space" using an out of autoclave epoxy reinforced with HM carbon fiber (Fig. 8).
- The structure was instrumented with four optical fibers, each one having nine FBGs. FBGs were bonded at the bars of the central section.
- The structural response of isogrid structures is rather complex. Because of slight manufacturing imperfections, the strain distribution was not uniform, as shown in Fig. 9, where sensors located at near identical locations show differences of 20%. Also, it behaves in a nonlinear way, changing from the loading to the unloading stage.
- A maximum compressive load of 330 kN was reached before the sudden failure of the structure. Some bars and nodes broke at the central region of the cylinder. Since there were many structural elements, load was redistributed through the structure, and the load capability was maintained.

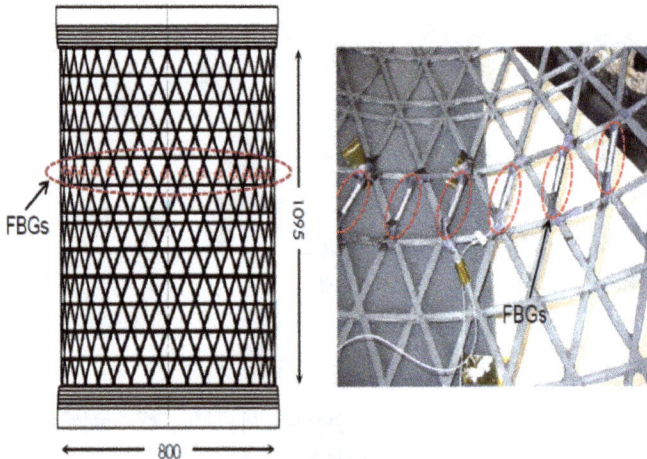

Figure 8.   CFRP isogrid structure. Detail of bonded FBG sensors.

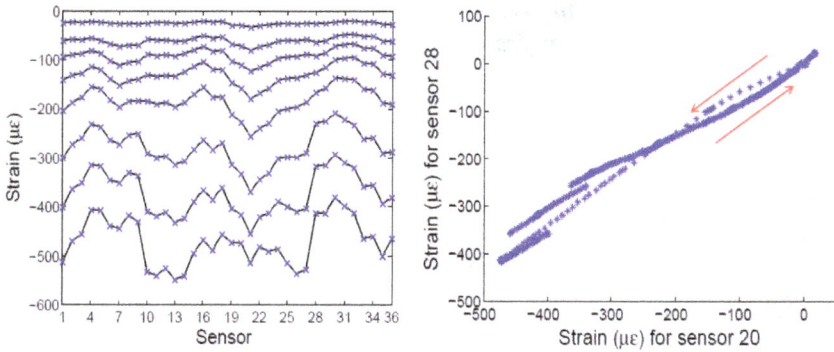

Figure 9.　Strain measurements at the central section. Loading/unloading cycle.

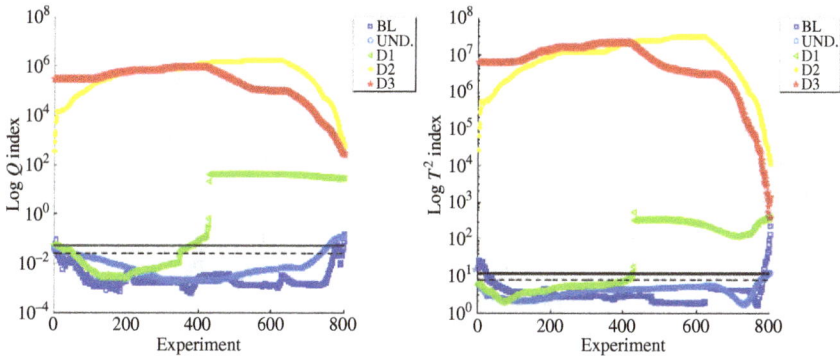

Figure 10.　Damage indexes for the isogrid structure under increasing compressive loads.

- The PCA algorithm was applied to these data sets, before failure of any structural elements (blue lines), and after failure of one of these bars, which is easily identified as the sudden step at the green line; subsequent failures produce larger changes at the damage index (Fig. 10).

- The experiment on the isogrid structure demonstrates that the failure of one bar was immediately detected, even if it did not cause a catastrophic failure, because of the high structural redundancy intrinsic to these kinds of structures.

3.3. *Conclusions from these Experiments Concerning Damage Detection from Strain Measurements*

- The high sensitivity of the PCA technique for the isogrid structure may be explained because for this experiment, the loads were monotonic (always compressive loads, of increasing levels), and because all the load paths were monitored (all the bars at the central section were instrumented).

- The experiment on the wind turbine blade shows a higher complexity, because several different failure modes may occur (debonding, cracks, etc.), and there may be different load distributions. The most influential factor on the strains are the external loads; consequently, these will always be the first principal components. The experiments combined different load positions; the number of sensors was considered realistic for the applications (four optical fibers, each fiber with several sensors, which have a distance of 2 m among them). Looking for the damage detection capability, it was found that a trailing edge debonding, a typical damage for wind turbine blades, is over the threshold of detectability if the debonded length was higher than 100 mm, a rather long crack from an aeronautic perspective, but still acceptable for wind turbine maintenance operations. The reason behind this "poor" sensitivity is that trailing edge debonding changes the general strain field paths and distributions minimally. Of course, partial cuts of the beam change the strain field more strongly, and consequently are easily detected.

4. Damage Detection by Distributed Sensing

The former procedure requires the structure to be tested under different load conditions to create large strain data sets that may be analyzed. Nothing can be said about the existence of damage by strain mapping if the structure is at rest, at an unloading condition. Sometimes SHM is intended to be done while the structure is working (for example, for wind turbine blades during normal operating conditions), but sometimes the structure needs to be inspected during overnight or maintenance stops, without any applied load. For these cases, the distributed sensing offers a possibility. The basic principle is simple; if an optical fiber is bonded or embedded with the structure, the OFDR may keep these readings as the reference for the undamaged, unloaded structure. Later on, in case a local damage has happened, and again the strain readings along the fiber is acquired, at the same unloading conditions as before, any strain data that appears at any point would mean residual strains caused by a damage at that point.

It has been shown in Fig. 4 that a delamination in a composite material causes, the relaxation of strains locally, which may be detected, as far as our sensor is located there. The approach proposed in Fig. 4, covering the whole surface by an optical Fiber that follows a crooked path, is feasible but unpractical, as it would require a very large length of fiber and current sensing length of about 80 m. But what seems useful, and now applicable, is to use this technology to survey critical areas, like at the edges of laminates around doors and man-holes, which have a higher risk of damage. By bonding the optical fiber a few millimeters away from the edge, a permanent inspection system of the perimeter is implemented, which can be interrogated anytime, without ambiguity and in a very fast process, just connecting the optical fiber end and doing one reading. All the fiber is interrogated at the same time, even at hidden positions.

Figure 11 illustrates the strains readings acquired by optical fibers bonded near the edge of a laminate (at distance 5, 10, 15 and 20 mm), after impacts were given at the edge, and the ultrasonic C-scan for the same plate after impact damage.

It was found that light impacts that produce very small delaminations at the ultrasonic C-scan, less than 2 mm deep, were not detected by the optical fibers, but the strong impact, with a higher internal delaminated area, was easily identified. (It's worth a mention that at this artificial colors image, that most of the figure in green = 0 microstrains, with a red spot (+150 microstrains) surrounded by a blue region (−100 microstrains).)

5. Monitoring Local Damages by Spectral Distortion

Most of the aeronautical structures are built as stiffened panels and shells, usually by a cobonding process for composite materials (skin is uncured, stiffeners are added as cured rigid elements). This is an efficient way to achieve a very high level of structural integration, but still one issue is unsolved: the stringer run-away or debonding starting at the tip. A classical way to solve this problem is by increasing the foot size (which makes the manufacturing of the stiffener more expensive), or by "chicken rivets". Nevertheless, it needs to be inspected in service, so an automated procedure to check the structural integrity is highly desirable.

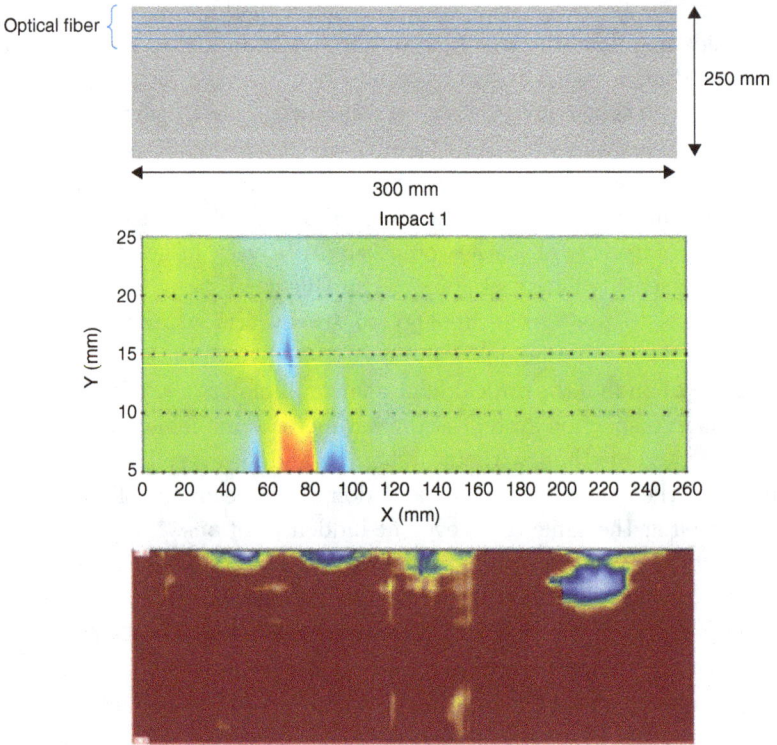

Figure 11.   Damage detection at laminate edge by distributed sensing Upper image:
Sketch of the position of the optical fibers.
Intermediate: Residual strains after impact.
Lower image: C-Scan of the laminate after impact.

Distributed sensing, as explained previously, may be useful to detect
the onset of stringer debonding. Just by including a plain optical fiber at
the adhesive line before curing, or bonded onto the stringer, information can
be obtained about the strong strains changes that happen at the debonded
region.

   An alternative to it would be to embed an optical fiber near the tip
with one FBG engraved at its tip, as shown in the next experiment.

   A flat laminate was prepared, the optical fiber with the FBG sensor
located on the adhesive film, the stiffener was added and the whole set was
cured (Fig. 12). Peeling loads, starting from the tip, grew a debonding. On
the left side of Fig. 13, the spectrum from this embedded sensor is shown,
before debonding; the strong peak splitting caused by the cure transverse

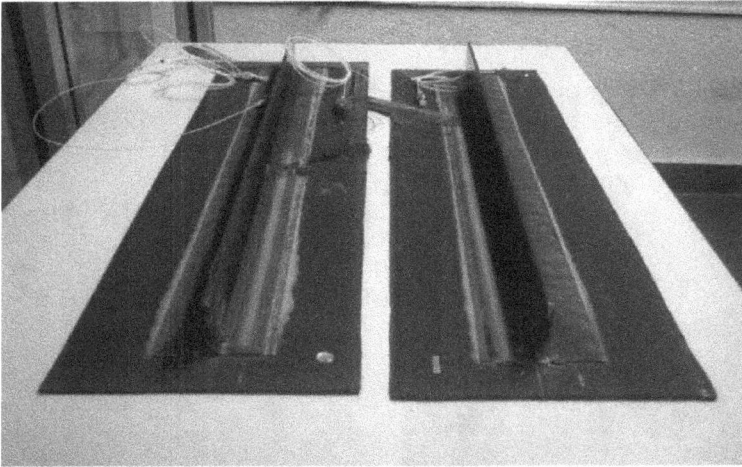

Figure 12. Manufacturing CFRP laminates with embedded sensors by cobonding.

residual stresses is remarkable. At the upper side of the same figure, the ultrasonic C-scan image, for this condition is shown.

On the right size of the figure, same results are shown, after the debonding crack has reached the sensor position; it is clearly seen how the spectrum changes and the transverse stresses disappear. The technique shows a high reliability, but only to detect damages that provoke a strain change at this specific point.

6. Model Assisted Probability of Detection (MAPOD)

From a user point of view, dealing with structural integrity and reliability of the inspection methods, the concept of probability of detection (POD) deals with the uncertainty associated with each inspection procedure, uncertainty coming from the sensitivity of the technique, crack shape, operator experience and environmental effects. The POD was developed to provide a quantitative assessment of NDT detection capabilities and focused on the smallest detectable flaw.

Just as an example, Fig. 14 (from Ref. [10]) shows the POD for internal cracks in metallic structures with several conventional NDT techniques. Damage tolerance criteria and scheduled maintenance may be programmed after this information, when damage growth laws are known.

It is found that the POD is related to the sensitivity of each specific NDT technique, but assuming adequate inspection conditions: qualified

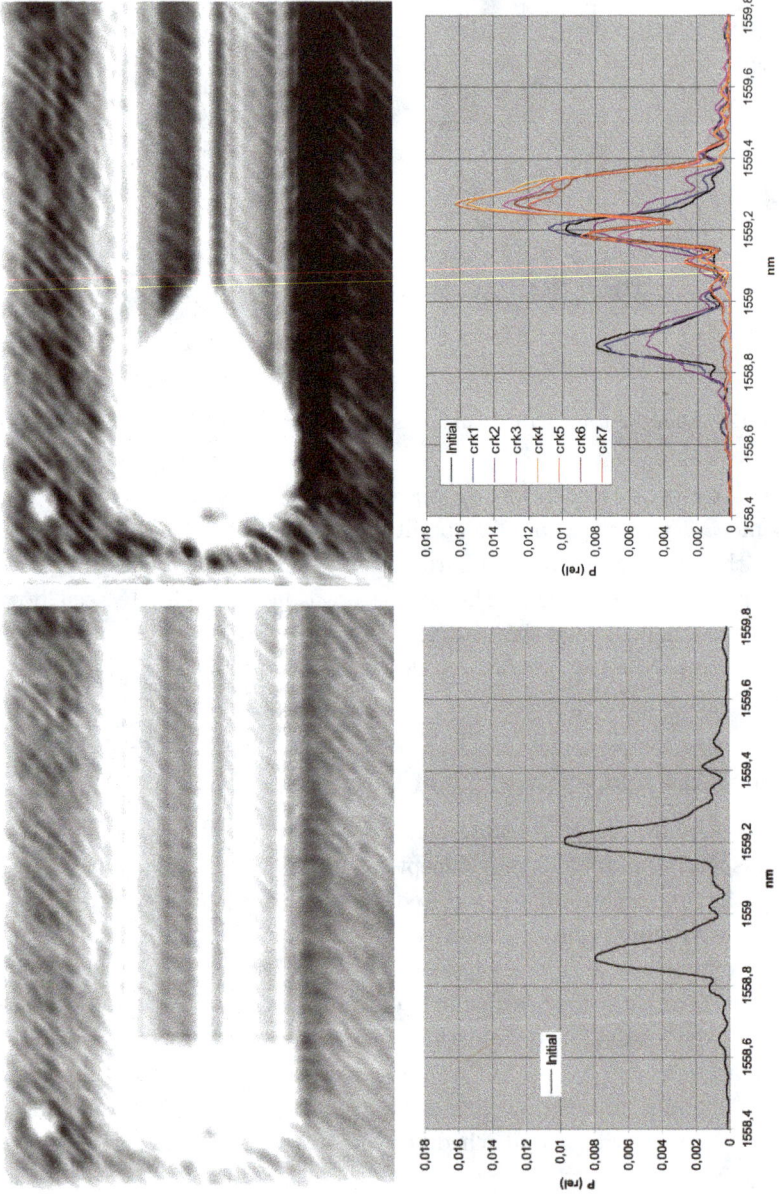

Figure 13. CFRP stiffened panel. Top: C-Scan images; Bottom: FBG spectrum.

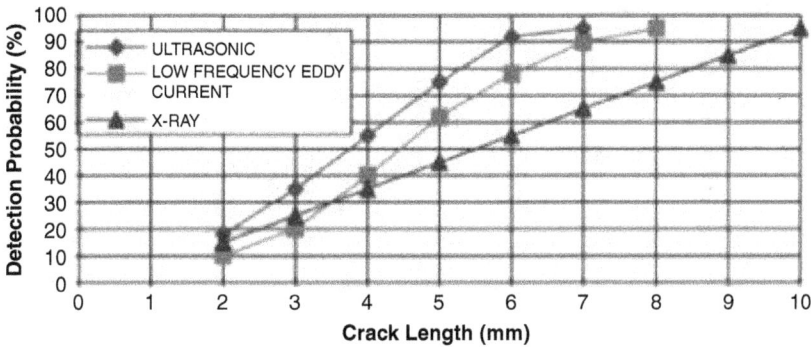

Figure 14. POD for internal defects (from Ref. [10]).

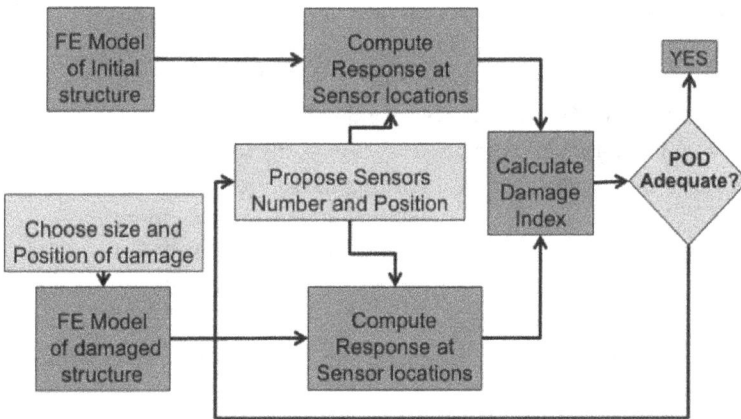

Figure 15. Model assisted POD for an SHM system.

inspectors, the sensing probe moves on the surface to inspect the whole region.

There is a need to develop a similar concept for SHM, if it has to be applied to assess the structural integrity of high-responsibility structures. But creating the set of experiments needed to quantify the statistical basis for the POD is much more complex than for conventional NDT. Charts similar to those given in Fig. 14, which are valid for every structure made with steel plates, are no longer possible. Each SHM system will be specific, not only for the technology used, but also on how it is deployed for each particular structure (number and position of sensors, and so on).

A conceptual approach to deal with this issue is proposed in Fig. 15.

7. Conclusions and Final Remarks

The potential of SHM technologies are huge, but they will not be applied as commercial off the shelf (COTS) products. They will always require a customized development. After 20 years of research and lab trials, the first industrial applications start to be implemented. As it has been pointed out at Ref. 11, it is important to distinguish between global and local applications, where for the latter case, the region of interest is smaller and well identified, leading to much simpler problems.

It is worthy to mention that in the same document Airbus point of view has been presented as comparative vacuum monitoring (CVM) and acousto-ultrasonics (AU) currently being the most mature SHM technologies. Nevertheless, fiber optic sensors promise a great potential, particularly for SHM applications for the whole structure, and Airbus has recently concluded a five-year research project (Japanese SHM Technologies for Aircraft Composite, JASTAC, structures) in close cooperation with Mitsubishi, Kawasaki, Fuji Heavy Industries, Japanese Aerospace Exploration Agency (JAXA) and the University of Tokyo; Results are presented at the EWSHM2016, and will be available at ndt.net. The concepts were quite similar to those presented in this chapter. Similar conclusions were attained at the review recently done by di Sante,[12] available as Open Access and which includes a long list of references.

There are some practical issues (like embeddable micro-connectors), and maybe some new conceptual approaches, before the technology can be considered mature and ready to be applied (TRL 9). Currently fiber optic sensors are probably TRL8 for strain measurements, but below TRL5 for damage detection purposes.

References

1. Meltz, G., W. W. Morey, and W. H. Glenn, Formation of Bragg gratings in optical fibers by a transverse holographic method. *Optics Letters* **14**(15) (1989) 823–825.
2. Froggatt, M. and J. Moore, Distributed measurement of static strain in an optical fiber with multiple Bragg gratings at nominally equal wavelengths. *Applied Optics* **37**(10) (1998) 1741–1746.
3. Giurgiutiu, V., *Structural Health Monitoring of Aerospace Composites*, 1st edition. Academic Press, USA, 2015.
4. Ostachowicz, W. and A. Guemes (Ed.), *New Trends in Structural Health Monitoring*, Springer, Germany, 2012.

5. Grabovac, I., T. Nuyens, and C. Davis, Packaging and mounting of in-fiber Bragg grating arrays for structural health monitoring of large structures, DSTO-TR-2490 (2010).

6. Measures, R. M., Smart structures with nerves of glass. *Progress in Aerospace Science* **26** (1989) 289–351.

7. Takeda *et al.*, Outline of the Japanese national project on structural health monitoring system for aircraft composite structures and JASTAC project, IWSHM2013, Stanford, USA, 2013.

8. Guo, H. *et al.*, Fiber optic sensors for structural health monitoring of air platforms. *Sensors* **11** (2011) 3687–3705.

9. Krisnhaswamy *et al.*, Low-cost adaptive array demodulation of Bragg-grating and interferometric sensors for health monitoring of marine structures, ONR-TR-2006403305 (2006).

10. Ratwani, M., Inspection technologies. At aging aircrafts fleets: structural and other subsystem aspects, NATO RTO EN-015/AVT-053 (2010).

11. Gardiner, G., Structural health monitoring: NDT-integrated aerostructures enter service, *Composite World*, August (2015).

12. Di Sante, R., Fiber optic sensors for structural health monitoring of aircraft composite structures: Recent advances and applications. *Sensors* **15**(8) (2015) 18666–18713.

Chapter 7

Impact Detection and Identification with Piezoceramic Sensors: Passive Sensing

Z. Sharif Khodaei* and M. H. Ferri Aliabadi[†]

Department of Aeronautics, Imperial College London, London SW7 2AZ, UK
z.sharifkhodaei@imperial.ac.uk
[†] *m.h.aliabadi@imperial.ac.uk*

This chapter presents structural health monitoring (SHM) methodologies for passive sensing, resulting in impact detection and identification. An overview of methods utilizing sensor data recorded by piezoelectric (PZT) sensors is presented. Particular attention is paid to data-driven methods using machine learning algorithms such as artificial neural network (ANN) and support vector machine (SVM) for the determination of impact location and energy. A range of experimental studies on the influence of operational conditions on the recorded sensor data are presented. Finally, a Bayesian-based optimization is presented for determining the optimal sensor configuration in complex structures. The optimization was applied considering uncertainty in the recorded data, probability of sensor failure and non-uniform probabilities of impact occurrence on the structure.

1. Introduction

Due to advances in composite material design resulting in superior properties, their application in the new generation aircraft (A350, B787) has increased significantly. However, composites are vulnerable to impacts causing both visible impact damage (VID) and barely visible impact damage (BVID), which if not detected on time, can lead to catastrophic failures. Uncertainty with respect to the likely energy of an impact event leads to conservative composite design and does not allow significant weight savings. Therefore, greater confidence from the knowledge of impact events can lead to less conservative design while still maintaining the no growth concept.

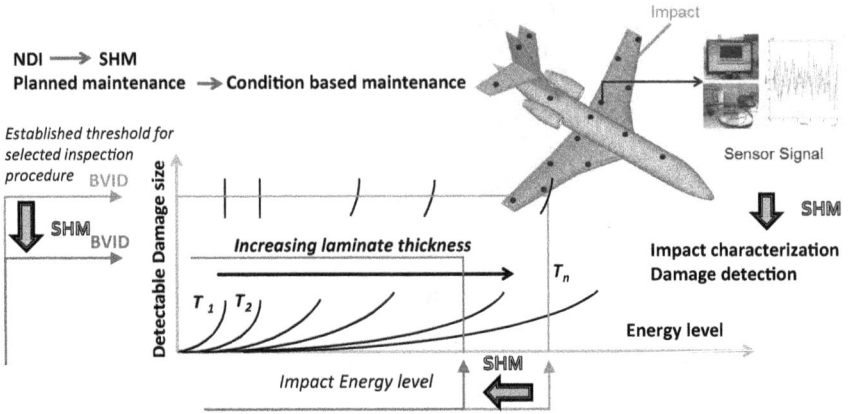

Figure 1.   Influence of SHM in composite design.

There has been a significant progress in the development of structural health monitoring (SHM) methodologies for impact and damage detection in the past two decades. Permanently installed sensors on the structure can monitor its condition during aircraft's operational life and detect any changes. Consequently, impact and/or damage can be detected and appropriate maintenance action carried out. The design strategy, for composite aircraft parts, currently follows the graph shown in Fig. 1. Depending on the probable impact energy levels for each part (horizontal axis) and the smallest detectable damage size based on the Non-destructive inspection (NDI) technology (vertical axis), the corresponding laminate thickness is chosen. The positive influence of SHM on the composite design is two-fold: first by reducing the detectable damage size in comparison to visual inspection, the laminate thickness can be reduced; second by having a better estimate of the impact energy levels by on-line monitoring of each part, a more accurate design could possibly reduce the laminate thickness and lead to significant weight savings. Currently, the probable impact energy levels are estimated by measuring the dent in metallic structures. The data gathered from the metallic aircrafts are used for the design of composites. However, this is not very accurate and can lead to overestimation in the design. The aim of the SHM technique applied to an aircraft is to provide a more realistic design scenario and avoid conservative design, resulting in weight savings, together with time savings by moving from planned maintenance to condition-based maintenance.

SHM techniques can be divided into two categories: impact detection and characterization and damage detection and identification. The first category is usually an outcome of a passive system where sensors capture the response of the structure due to an external event such as an impact. The transducers, in this case, are used as sensors only and by processing the measured response of the structure, impact can be detected and characterized. The second category is usually an outcome of active sensing where the structure is actively excited and its response is measured both in the pristine state and in current state where damage might be present. There are many active sensing methodologies developed for detecting and characterizing damage in metallic and composite structures (see, e.g., Refs. [1–5]). The focus of this chapter, however, is on impact detection and identification in composite structures. The structure of the chapter is as follows: Section 2 highlights the important factors that influence the response of a structure under real operational conditions, in particular on the recorded sensor data. Section 3 presents an overview of the existing methodologies for impact detection and characterization. Section 4 focuses on the two methods that are utilized for impact detection and characterization and describe them in detail. Section 5 presents an optimization strategy for optimal sensor number and location, which includes uncertainty in the data representing real operational conditions and sensor failure increasing the reliability of the impact detection methodology. The results the proposed methods on a composite stiffened panel are presented in Section 6 following experimental validation at coupon level.

2. Passive Sensing of Airframe Structures

When an impact event occurs on a sensorized structure, elastic waves are generated and after travelling along the surface of thin plates, they are recorded by permanently mounted sensors on the structure. The aim of SHM impact detection methodologies is to utilize the sensor data to localize and characterize impact events. Even though many passive sensing methods have been established for impact detection and characterization, many of them are limited to laboratory environments, isotropic material, simple plates with low velocity and energy levels. All of the developed methodologies are based on the data recorded during an impact event by permanently mounted sensors on the structure. To develop an impact detection and identification methodology for an aeronautical

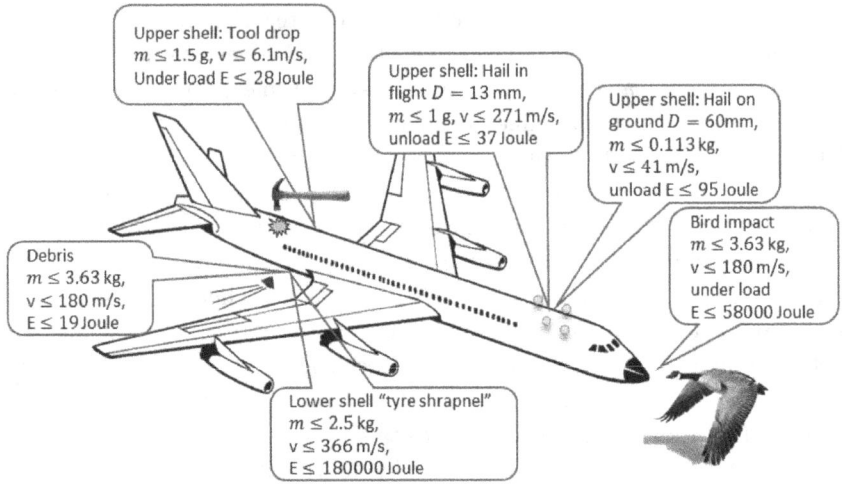

Figure 2.   Possible impact scenarios on an aircraft.

application, environmental variability and impact energy variations needs to be taken into account. In this section, some of the most important factors that influence the response of a sensorized structure on an impact event, under real operational conditions, are presented. In particular, the effect of operational conditions on the recorded data is highlighted with experimental sensor data recorded by impacting composite coupons.

2.1. Impact Characterization

The probable impact scenarios on an aircraft are much more complex than limited number of tests carried out on coupons in controlled laboratory environments. Each part of the aircraft is prone to different level of impact energy (see Fig. 2) and probability of impact occurrence (Table 1). One categorization of impact is based on their energy level: low-energy and high-energy impacts. However, impact energy alone is not a good criterion for classifying the impact as two impacts of the same energy but different mass and velocity combinations can trigger a completely different response in the structure. An impact of 6 J energy, with small mass and high velocity, can result in damage in a composite plate whereas another impact of 6 J energy but with large mass and small velocity will not cause any damage in the structure.[6] It is more appropriate to characterize impacts with respect to the contact force.

Table 1. Global percentage of impacts by zones on an aircraft.[7]

Zones	Fuselage sections (%)	Wings (%)	Nose (%)	Cone and rear fuselage (%)	Doors (%)	Passenger door surrounds (%)	Cargo door surrounds (%)
Impact	7	13	7	5	15	31	22

Based on the impactors, the impact is divided into small mass (e.g., debris) and large mass (e.g., tool drop) where the response of the structure to each type can be very different. One of the main differences between the small mass impact and large mass impact is in terms of impact duration; see Fig. 3 where small mass impacts have much shorter duration. Therefore, to know whether the occurred impact is alarming or not (could have initiated damage), it is more reliable to detect the contact force, and not the energy. The delamination threshold force of a composite plate can be determined using the following relation[8]:

$$F_d = \pi\sqrt{16D^*G_{\text{IIc}}} \tag{1}$$

In this equation, G_{IIc} is the mode II interlaminar toughness and D^* is the effective plate stiffness approximated by

$$D^* = \sqrt{D_{11}D_{22}(A+1)/2}; \quad A = (D_{12} + 2D_{66})/\sqrt{D_{11}D_{22}}. \tag{2}$$

In addition, impacts can occur from inside and outside of the structure where both the contact force and the response of the recorded sensors can be very different, see Figs. 4 and 5.

Therefore, it is important that when developing a meta-model for impact detection and identification for a real structure, all possible impact scenarios are taken into account.

2.2. Impact Categorization

It is desirable for the SHM system to be able to distinguish between the impacts of very low energy and those which may have caused damage in the structure. In impact detection, false alarm can be defined as identifying an impact event as alarming when the level of the energy is low enough as not to cause any harm to the structure. If an impact of very low energy is identified as alarming, it can cause significant monetary loss due to unnecessary maintenance and the grounding of the aircraft. It

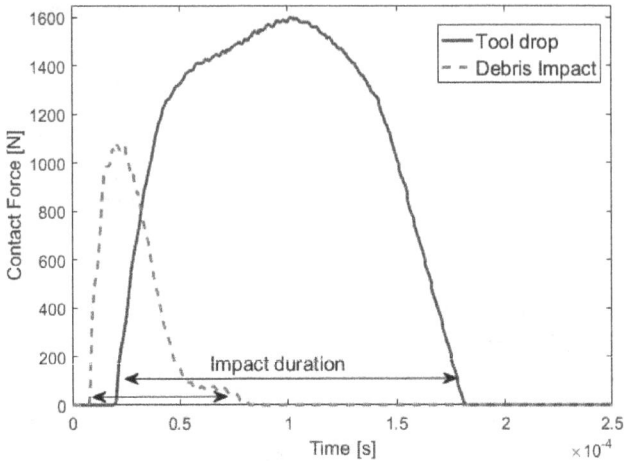

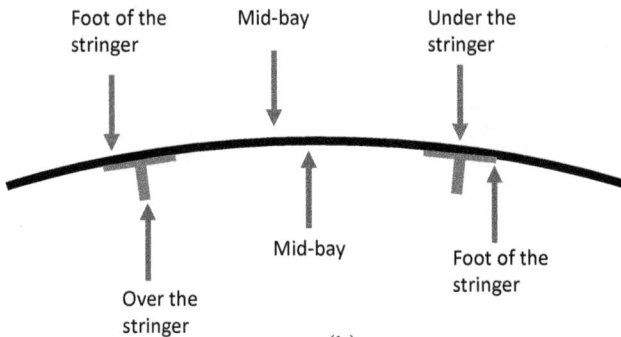

Figure 3.   (a) Contact force for small mass versus large mass impacts; (b) impact location.

is therefore important that the false alarm rate, i.e., the percentage of non-damage-causing impacts (false impacts) that are mistakenly identified as damage-causing impacts (true impacts), is minimized. Impacts of different energy will produce different responses in the structure. Therefore, to categorize impacts, features extracted from the sensor signals can be used such as the frequency content,[9] relative energy or the power spectral density (SPD) of the captured signals.[10]

A detailed study was carried out in Ref. [10] to propose a robust methodology of categorizing impacts. Impacts of different mass and velocity

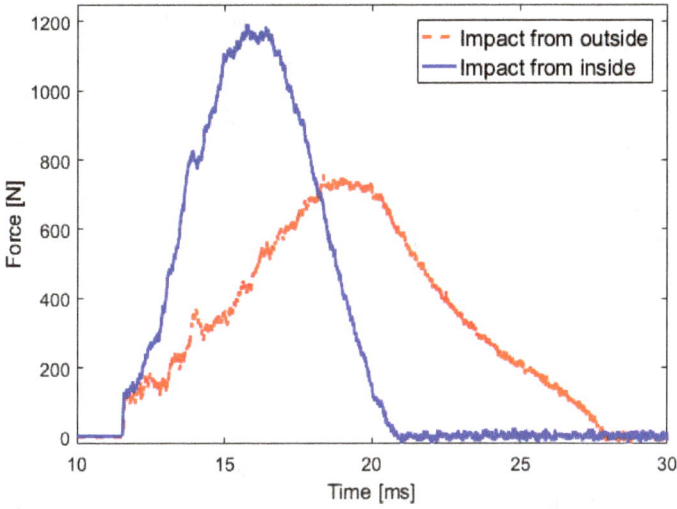

Figure 4. Contact forces of the same impact, one occurring from inside and one from outside of a curved composite panel.

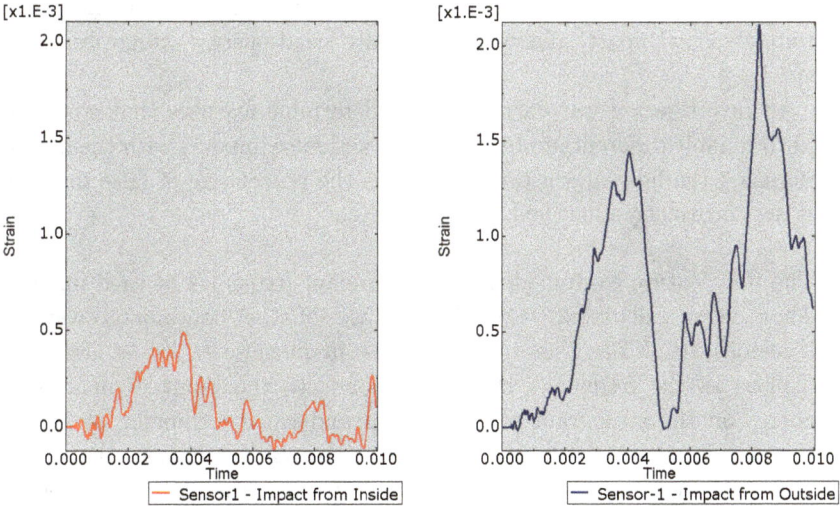

Figure 5. Sensor signals recorded due to an Impact from inside and outside of a curved composite panel.

Table 2.   Mean and standard deviations of the average instantaneous frequencies for each impactor.

Energy level	Impact mass/Energy	Mean instantaneous frequency (Hz)	Standard deviation of instantaneous frequency (Hz)
E0-1	2.7 g, 9.54 mJ	605.96	147.02
E0-2	12.7 g, 44.85 mJ	454.82	98.88
E1	24.46 g, 86.38 mJ	205.47	24.70
E2	24.46 g, 119.98 mJ	206.80	24.45
E3	31.60 g, 155.00 mJ	197.50	27.81
E4	38.74 g, 190.02 mJ	188.01	25.81

were conducted on a composite plate and the response was recorded by eight surface-mounted piezoelectric (PZT) transducers. The combination of different masses and energies are summarized in Table 2. The two lightest impactors from the experiment were chosen as the false or "non-damaging" impacts and given the designations E0-1 and E0-2. The four heavier impactors were chosen to be the true or "damaging" impacts and were given designations E1–E4, each having impact energies that are about 35 mJ apart, allowing for equally sized energy categories to be created.

An investigation was carried out to determine features that could be used to reliably differentiate between true and false impacts with the aim of minimizing the false-alarm rate, as well as the percentage of false impacts that are incorrectly identified as true impacts.

- The first feature examined was the dominant frequency of each impact. They were calculated through the use of the continuous wavelet transform (CWT). The instantaneous frequency, f_i, of a signal is defined as the frequency that corresponds to the point of maximum energy in the scalogram of that signal and can be thought of as the signal's dominant frequency. The instantaneous frequencies may vary substantially between signals for a single impact; therefore, the average instantaneous frequency, $\bar{f}$, of an impact might be a more effective identifier. The average instantaneous frequency (Hz) for each of the 64 impact locations for each energy level were calculated and plotted in Fig. 6, which shows a clear difference between the false impacts and the true impacts.

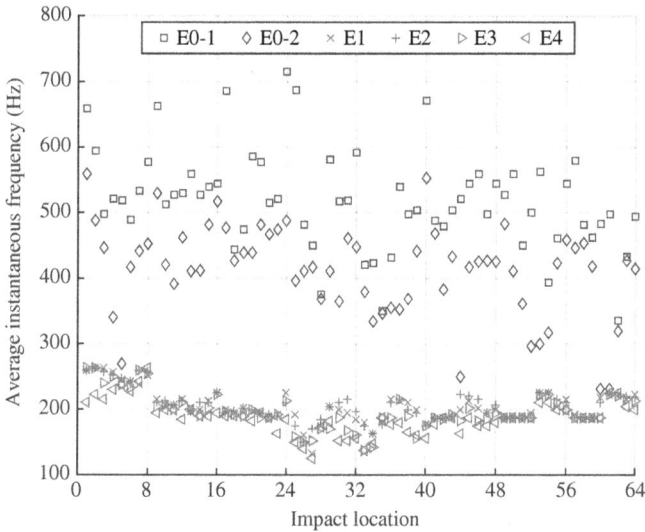

Figure 6.   Average instantaneous frequency (Hz).

The frequencies of energy levels E1–E4 are concentrated between 100 and 300 Hz while the frequencies of E0 are distributed over a wider range from 200 to 800 Hz.

- The second feature involved calculating the relative energies of the impacts. This was also achieved through the use of CWT. The scalogram of a signal is obtained by calculating the squared modulus of the CWT coefficient at each scale a and time b. By calculating the squared modulus of the CWT coefficients at $\bar{f}$ for each time b and for each of the eight sensor signals, and then numerically integrating the results over time b, a measure of the energy content of each signal $Q(s)$ can be obtained, see Fig. 7.

- The third feature that was investigated involved calculating the power spectral density (PSD) of the captured signals, providing a measure of the power of the signals, which could be related to the impact energy. The PSD of a signal describes how the power of that signal is distributed over the frequency domain. Its units are power per frequency (W/Hz) and by integrating PSD over a frequency range, it is possible to obtain the power within that frequency range. PSD was evaluated in the frequency range 0–250 kHz adopting Welch's PSD estimate method. From the result shown in Fig. 8, it can be seen that there is a clear distinction between

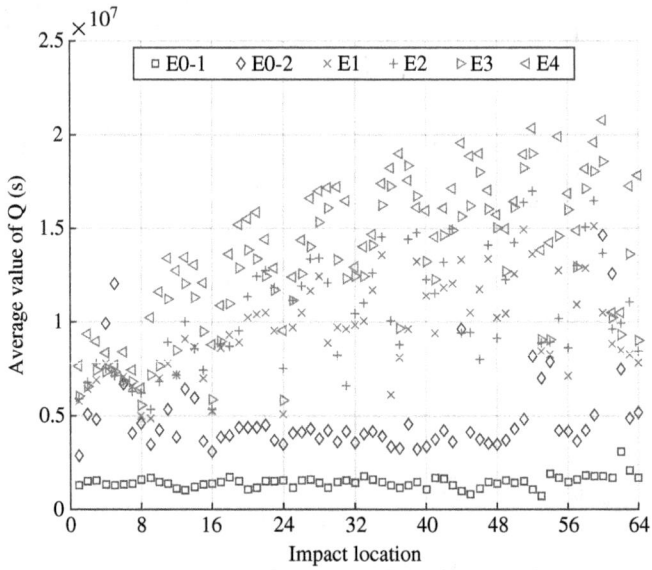

Figure 7.   Average value of $Q(s)$.

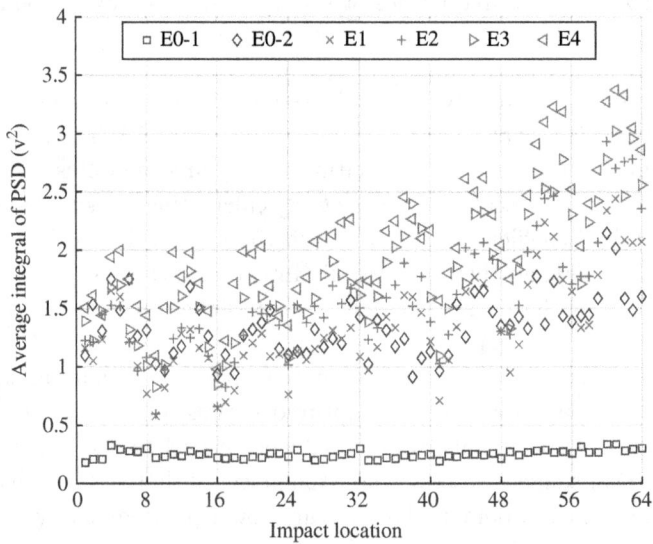

Figure 8.   Average integral of PSD (v2) per impact location.

E0-1 and E1–E4, which is to be expected due to the fact that the impact energy of E0-1 is much less than that of E1–E4. However, it was not expected that E0-2 would have similar PSD integral values to E1 and E2.

When comparing all three features presented in Figs. 6–8, it is clear that the PSD integrals are the least effective at differentiating between true and false impacts, with the most effective feature to be the average instantaneous frequencies. This classification strategy is presented for a fairly idealized case involving impacts of a relatively small range of masses, impact energies and angles of attack, and should therefore be seen as the first step in proof of concept for impact classification to structures under real operational conditions.

2.3. *Effect of Environmental Factors on Impact Response*

Impact on an aircraft can occur under different operational conditions. One of the most important varying factors is the temperature. The question is whether the recorded sensor data used for impact detection and identification is influenced by the temperature variation and if yes, will this variation affect the results. The influence of temperature on impact generated signals was investigated in Ref. [11].

An impact experiment on a composite panel, with eight surface-mounted PZT sensors (denoted by circles in Fig. 9), was repeated at five different temperatures: RT: room temperature $(25°C)$, RT $+$ $10°C(35°C)$, RT $+ 20°C(45°C)$, RT $+ 30°C(55°C)$ and RT $+ 45°C(75°C)$. The panel is made of M21 T800S prepreg with the following layup $[0/+45/-45/90/0/+45/-45/90]_s$. The panel was impacted at 18 different locations (see the crosses numbered in Fig. 9). The sensor signal recorded in sensor 1 (see Fig. 9) for a fixed location of impact has been compared for varying temperature in Fig. 10. The times of arrival of the signals are slightly shifted due to the temperature variation as shown in Det. 2 of Fig. 10; however, this shift is within the noise level in the data. The change in amplitude, however, is more significant (Det. 1 of Fig. 10). This means that if the amplitude of the signal is used in impact identification, the temperature effects should be compensated for, otherwise it will increase the error of the prediction.

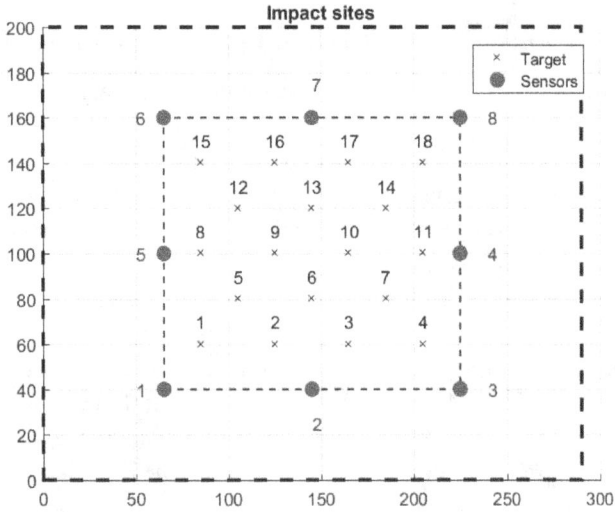

Figure 9.   Experimental setup for impacts at varying operational conditions.

2.4. *Angle of Impact, Shape of Impactor, Mass of Impactor*

The variation of the generated guided wave due to impacts of different mass and velocity combinations has been reported in Ref. [9]. However, the impactor shape and angle remained constant. Impacts that can occur in real operational conditions consists of various impactor shapes, i.e., they can have sharp angles and are not always hemispherical. In addition, impacts are not necessarily perpendicular to the structure and they can hit the

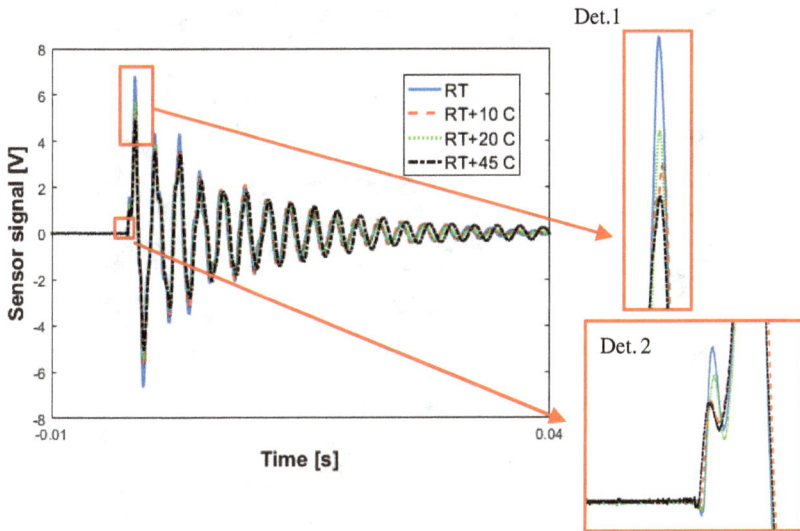

Figure 10. Effect of temperature on the signal at Sensor 1.

panel at different angles. The question that remains is whether, for the development of a meta-model to detect and characterize impacts, the angle and shape of impactor should be taken into account if they have an influence on the recorded sensor signal. To answer this question, an experiment was set up by[11] to record sensor data generated by impacts on a horizontal and inclined composite panel of 10° and 20° as shown in Fig. 11. The sensor signals recorded by the same sensors and the same impactor but generated due to different impacting angle has been compared in Fig. 12. It can be seen that not only the time of arrival (ToA) of the signals are affected by the impact angle (even when it is as low as 10°) but also the amplitude of the recorded signal is changed. This concludes that when ToA is used in developing a meta-model for impact detection, the variability in the impact angle, which will occur in real operational condition needs to be considered.

In addition, the influence of different impactor shape has also been investigated in Ref. [11]. Two impactors of similar mass were chosen: a spherical impactor and a cubic impactor. The impact height for both was chosen to be the same i.e., 36 cm (resulting in similar impact energy) and the impacts were carried out on the flat composite panel covering all 18 impact sites as indicated in Fig. 9.

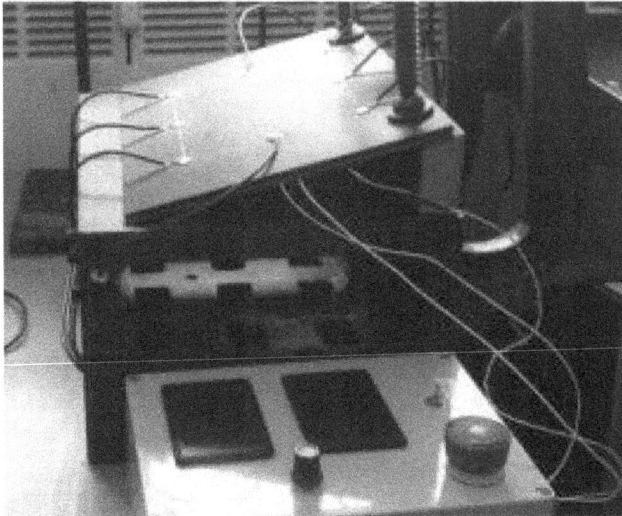

Figure 11.   Impact test on an inclined composite panel.

The results shown in Fig. 13 indicate that there is slight change in the ToA of the signal between the cubic and spherical impactor but this can be considered within the noise threshold. However, the change in amplitude of the signal is significant, which indicates that two impactors of similar mass but different shape can induce a different response in the structure. Therefore, in terms of locating the impact, the ToA information can be used for any impactor shape for developing a meta-model but not the amplitude of the signal, which is usually linked to contact force reconstruction and consequently damage initiation.

The next test was to investigate the influence of different mass for the cubic impactor. Three different cubic impactors of mass 17.5, 9.0 and 2.2 g, respectively marked as Cube 1, Cube 2 and Cube 3 all impacted from the height of 36 cm. The sensor responses in Fig. 14 show no consistency in the ToA or amplitude with increase or decrease of mass. It can be concluded that for impactors of sharp edges, the response of the structure depends very much on the angle and side of the impactor. Therefore, impact characterization will be very difficult to be carried out. In addition, the results emphasize the importance of developing an algorithm, which is able to include the variations in the impactor shape, mass, velocity and angle; otherwise it will not be applicable to structures under real operational condition.

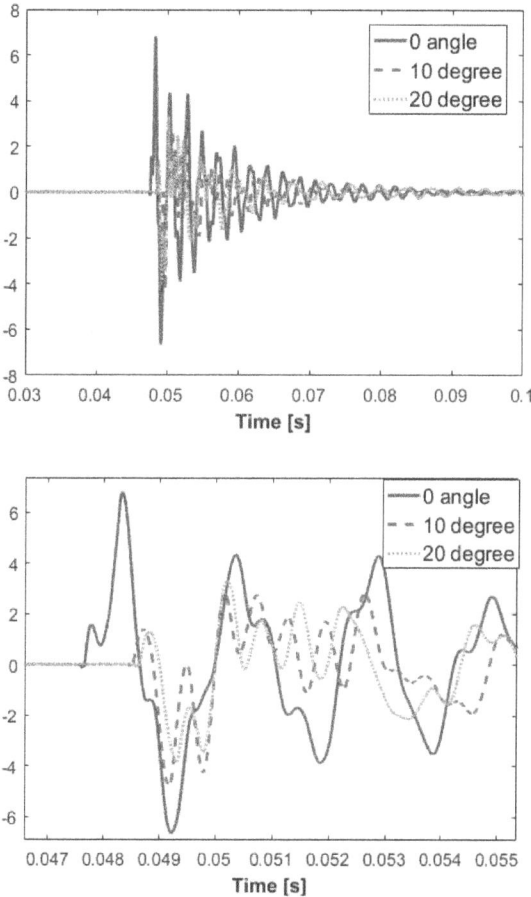

Figure 12.   Sensor signal recorded by the same sensors due to different impact angles but the same impactor and the same impact location.

3. Impact Detection and Characterization Methodologies

There are various impact detection methodologies based on sensor data, which have been developed over the past two decades. They can be grouped into three categories:

(a) **Trigonometric location techniques:** These include methods such as triangulation based on the ToA of the waves. The difference in arrival time of an acoustic–emission wave at a pair of sensors represents the difference in the distance of the emission source from the two sensors. The emission source location can then be represented by a

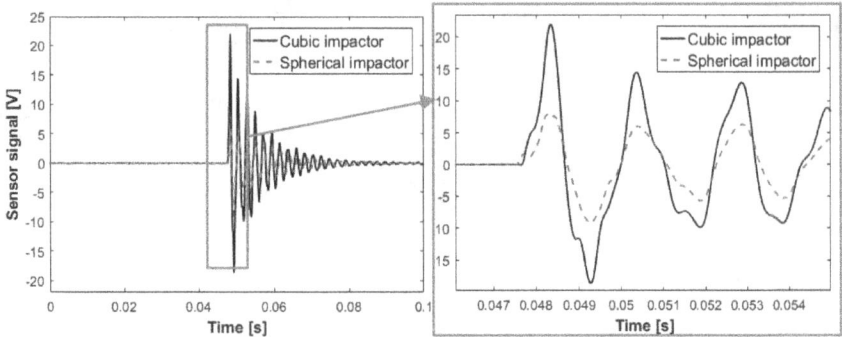

Figure 13.　Same impact energy, same location, but different shape of impactor.

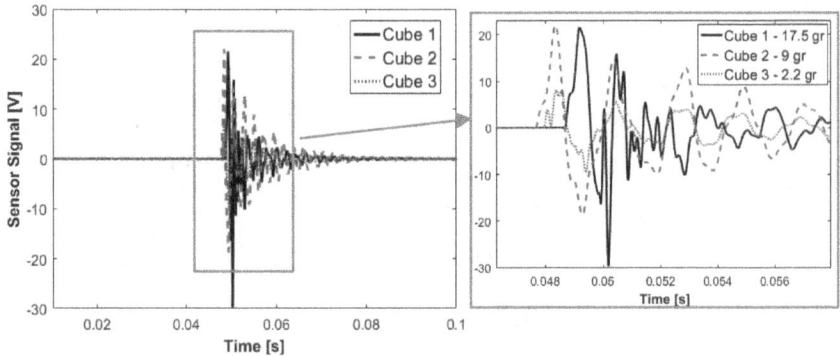

Figure 14.　Same impact location and height but different mass, cubic impactors.

hyperbola defined by the difference in the ToA with the two sensors as foci. By having an array of three sensors, the intersection of the pair of hyperbolae defines the location of the emission source. The equation for a pair of hyperbolae involve square root, which makes the analytical solution rather difficult. An alternative approach, however, can be employed where the acoustic emission (AE) source $P(x, y)$ is located at the point of intersection of the circles about sensors $S_0(0, 0)$, $S_1(x_1, y_1)$ and $S_2(x_2, y_2)$. If the source is located at distance r from S_0, the path differences δ_1 and δ_2 from sensors S_1 and S_2 can be denoted as $\delta_1 = PS_1 - PS_0 = t_1 v$, and $\delta_2 = PS_2 - PS_0 = t_2 v$, where t_1 and t_2 are differences in ToA measured for sensors $S_1 - S_0$ and $S_2 - S_0$, respectively and v is the velocity of the propagation in the material.

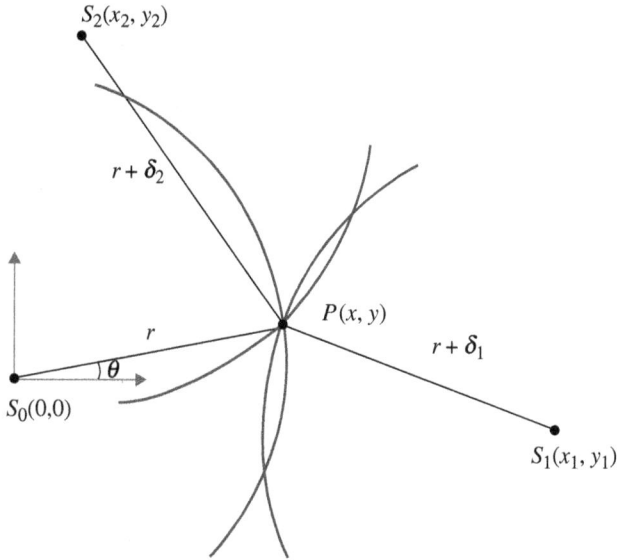

Figure 15.   Triangulation of an AE source based on three sensor network.

The AE source can then be located as the intersection of the circles centered at S_0, S_1 and S_2 with radii $r, r + \delta_1$ and $r + \delta_2$, respectively, see Fig. 15.

An analytic solution of the set of equations exists (see Ref. [12]). A value of the angle θ must yield positive r to be a valid solution. In most of the cases, there is only one valid solution of θ for a given pair (t_1, t_2). However, when two solutions of θ exist, the corresponding source location relates to two intersection points of hyperbolae. In order to resolve the ambiguous source location, a fourth sensor is necessary.[12]

In isotropic structures the velocity of the wave is constant in all directions, and triangulation is achieved with three sensors. For anisotropic material, it is possible to obtain velocity characteristics as function of the angle of the wave propagation to estimate the location of the impact combined with triangulation techniques.[13] In the case of anisotropic materials, the strain waves recorded by sensors S_1, S_2 and S_3 can assume three different impact positions at A_1, A_2 and A_3 (see Fig. 16). This results in three propagation paths $S_i A_i$ ($i = 1, 2, 3$) with variable angles α_i. For each angle, the unknown distance $S_i A_i$ is calculated as $S_i A_i = v_i t_i$ where v_i and t_i are the impact wave propagational velocities and arrival times. If the impact velocity is known as a function of

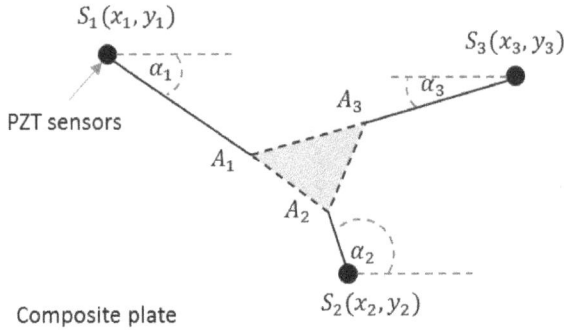

Figure 16.   Sensor triangulation in composite plates.

angle, then an optimization procedure can be carried out to locate the impact.

By having a distributed sensor array in the structure, a hyperbolic positioning algorithm can be used to locate impacts in composite plates using six PZT transducers (see Fig. 17).[14] Locating an AE source $P(x,y)$ is an inverse problem. If the transducers are located at a distance $\delta_i (i = 1,\ldots,6)$ from the source, and d_{km} $(k = 1,3,5; m = 2,4,6)$ are the distances between each pair of transducers k and m, the coordinates of the AE source can be determined by solving the following set of nonlinear equations:

$$\begin{cases} \|\delta_i\|^2 = (x_i - x)^2 + (y_i - y)^2 \\ t_i = \dfrac{\|\delta_i\|}{V_{g,i}}, \end{cases} \tag{3}$$

where $V_{g,i}$ is the velocity of the propagating wave recorded by the ith sensor, t_i is the ToA of the wave and (x_i, y_i) are the coordinates of the ith sensor. However, to use this approach, it is required to reduce the number of unknown variables to the number of equations such that each pair of sensors will experience the same group velocity, i.e., placing each pair very close to each other. This is a limitation for complex structures such as fuselage panels where stiffeners and frames are present.

(b) **Model-based algorithm:** It involves modeling the dynamic response of the structure subjected to a known impact location.[15,16] Measured responses of sensor output are compared with the estimated response from the model. This is achieved through modeling the wave

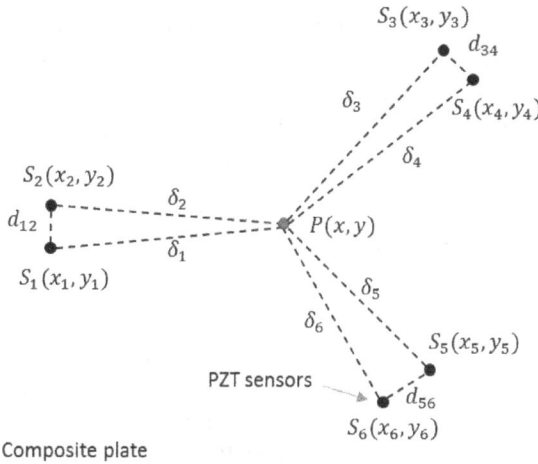

Figure 17.   Sensor arrangement for impact source localization in anisotropic composite.

propagation generated by impact via the equation of motion in plates and modeling the sensor response.

$$M\frac{\partial^2 w}{\partial t^2} + Kw = Ff(t), \tag{4}$$

where M, F and K are the mass, force and stiffness matrices respectively, t is the time, $w(x, y, t)$ is the displacement and $f(t)$ is the force history. Through an iterative procedure, transfer functions are developed to estimate the system response to an impact event and to locate the impact. Model-based algorithms can be used to both locate the impact and to reconstruct the impact force history from the model response. One method to build a transfer function between the impact force and the sensor data is to take advantage of the convolution integral.[17] In this method, it is assumed that the dynamic system (s) is linear time invariant (LTI), which can be expressed by a time convolution integral:

$$u(t) = \int_0^t s(t - \tau)f(t)d\tau, \tag{5}$$

where $f(t)$ is the impact force history and $u(t)$ is the measure response i.e., the recorded sensor data. In the frequency domain the convolution equation becomes

$$U(\omega_n) = S(\omega_n)F(\omega_n), \tag{6}$$

where U, S and F in Eq. (6) indicate the frequency domain representation of the displacement, the LTI system and impact force respectively. The circular frequency is represented by $\omega_n = 2\pi(n-1)/N\Delta t$, where N is the total number of data points and Δt is the sampling time step. Given frequency domain representations of the system and the sensor signals, the components of the spectral force can be obtained from

$$F(\omega_n) = U(\omega_n)/S(\omega_n). \tag{7}$$

An inverse Fourier transform of the spectral force will provide the reconstructed force in the time domain. The limitation of most model-based methods is that they are restricted to linear cases, which implies that the impact energy has to be small enough not to cause large deflections (see Fig. 18 and Ref. [9]). However, aircraft panels are very likely to undergo large deflection when they are exposed to impacts during their service life. Therefore, the drawback of model-based approaches is their application to complex structures.

(c) **Machine learning algorithms (data-driven techniques)**: This group of techniques require modeling complex relationships (meta-model) between input and output data. Artificial neural networks (ANNs) are mathematical models that can model complex nonlinear relationship between inputs and outputs. They can be applied to localize impact and reconstruct force history based on features extracted from sensor data.[18–20] ANN is a machine learning algorithm that adapts its structure (weights) based on the input and output of the system during the learning phase through chosen training functions. A large set of input and output data is required to reach convergence. ANN technique has been successfully applied to both impact localization and force reconstruction.

Despite having many advantages, the feedforward network has few fundamental flaws. It has been recognized that there are no set of rules to determine the optimal ANN architecture in terms of parameters such as number of required input data, hidden layers, neurons in each layer and the type of training function. In addition, the existence of many local minima can stop the training process to reach generalization. Alternatively, a new class of neural networks called support vector machines (SVMs) is proposed to model non-linear system.[21] SVM solutions are characterized

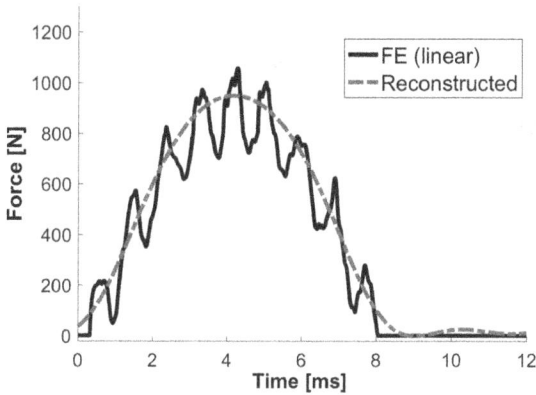

(a)

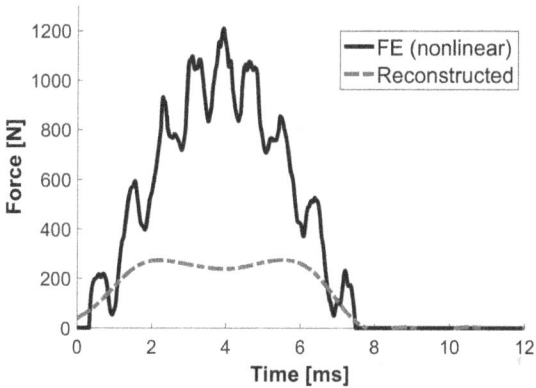

(b)

Figure 18.   Force reconstruction in a composite plate.

by convex optimization problems, which determines the architecture of
the SVM model automatically, unlike ANN. SVM has shown to reach
regularization with much less data in comparison to ANN.[22,23] This
is a very attractive feature as having a large data base of all impact
scenarios for each structure is very expensive. Another method to reduce to
computational size of the problem is to combine ANN with trigonometric
location techniques such as trilateration to reduce the computational
cost to one-tenth as proposed by De Stefano *et al.*[24] ANN and SVM
methods for impact detection and characterization will be explored in more
detail next.

4. Machine Learning Algorithms

Two classes of machine learning algorithms that have been applied to passive sensing, i.e., impact detection and characterization, are ANN and SVM. Each of these techniques is further explained in the following sections.

4.1. *ANN*

ANN is a machine learning algorithm that uses a set of training data to build a non-linear function relating a set of input data to output data. For passive sensing, the inputs to the ANN are features extracted from sensor signals and the outputs are impact locations or force magnitude (see Fig. 19). If there are too many parameters in the model compared to the number of training patterns, there might be the problem of overfitting. One way to ensure generalization is to use enough data for training each network weight or to use an early stopping regularization. Another way to avoid the overfitting is to evaluate the performance of the network on an independent test set. In this case the process of establishing a neural network can be divided into three phases: training, validation and testing.

Figure 20 presents the simplest kind of neural network which is a single layer perceptron network: the inputs are directly fed into the output via series of weights. The product of weights and inputs are summed and passed onto an activation function which defined the output of the node. The perceptrons can be trained by simple learning algorithms.

- During the training phase, the weights are adapted. The method that is adopted in this work is the backpropagation learning rule, which starts

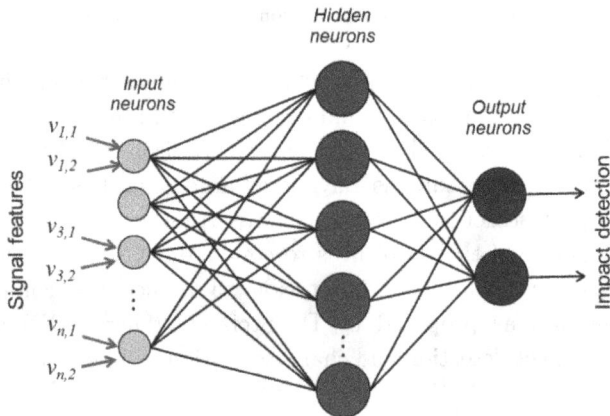

Figure 19.   Example of an ANN architecture.

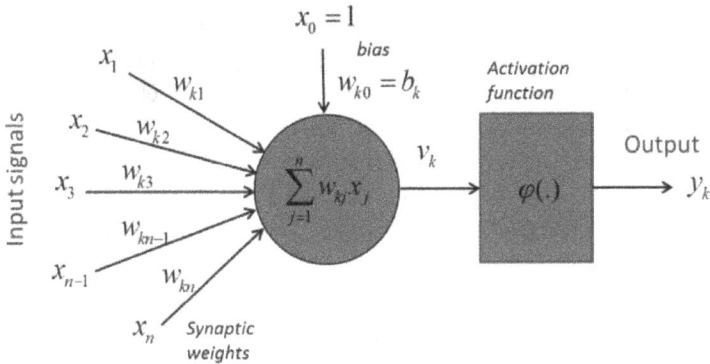

Figure 20.   Single layer feedforward perceptron.

with a set of random weights and repeatedly adjust the weights until the error converges to a set limit. To adjust the weights properly, one applies a general method for nonlinear optimization called the gradient descent.

- The validation phase is used for optimizing the network structure (number of hidden layers, number of neurons per layer, input features, etc.). Unfortunately, there are no sets of rules to determine an appropriate network structure. Therefore, different network structures should be examined in a case study to find its optimum structure.[25]

- The test phase is used at the end to assess the generalization and the performance of the network. Generalization of the ANN is its ability to handle unseen data. This assessment is done by the error measured on an independent set of data (the data set, which has not been introduced to the ANN during training). Therefore, the error measure can define the performance of the established ANN.

For an SHM system to be employed reliably in practice, it should be capable of distinguishing between different impact events, which may result in a damage or not (false alarm) as well as operating with a high reliability under operational conditions as well as the case when one or more sensors have become faulty during service. One way to increase the reliability of any decision-making algorithm is by adopting Bayesian updating[26–28] or Kalman filter.[29–31]

4.2. *LSVM*

Another class of neural networks is SVM, which is suitable for modeling nonlinear systems.[21] Unlike ANN, the architecture of a SVM model is determined automatically by solving the convex optimization problem.[32] The superiority of LS-SVM over MLPs is that it requires less data for training and in some cases it was shown to have a better accuracy.[33]

Several studies of impact identification using LS-SVM have been carried out[23,34,35] each with different levels of complexity in terms of impact scenarios and geometrical complexity of the structure under impact. LS-SVM model has the following representation in primal weight space:

$$Y(X) = \omega^{\mathrm{T}} \varphi(X) + b, \tag{8}$$

where X and Y represents the input and output vectors respectively, $\varphi(\cdot)$ is a nonlinear function mapping the input pace to a feature space and ω and b are the weight and the bias vectors. Given a data set $\{x_k, y_k\}_{k=1}^{N}$, the following optimization problem can be formulated in the input weight space:

$$\min_{w,b,e} \mathcal{J}(\omega, e) = \frac{1}{2} \omega^{\mathrm{T}} \omega + \gamma \frac{1}{2} \sum_{k=1}^{N} e_k^2, \tag{9}$$

where γ is regularization parameter and e is residual error. Such a minimization problem can be solved by constructing the Lagrangian and deriving the dual problem. For more details, see Ref. [32] LS-SVM uses kernel-based learning methods for pattern analysis with the training data. This approach is called the "kernel trick" where a kernel function is used, which enables them to operate in a high-dimensional implicit feature space without ever computing the coordinates of the data in that space, but rather simply computing the inner products between the images of all pairs of data in the feature space. This operation is often computationally cheaper than the explicit computation of the coordinates.

The select in of kernel function has a significant influence on the model accuracy. There are two popular approached: (1) cross validation[23] and (2) Bayesian inference.[35,36] In a k-fold cross-validation, the data set is randomly divided into k equal-sized subsets. In the ith fold of the cross-validation procedure, the ith subset is used to estimate the generalization performance of a model trained by the remaining $k-1$ subset. The estimate of the generalization performance of a model trained by the entire training data is provided by its average performance of observed over all k folds. Generally

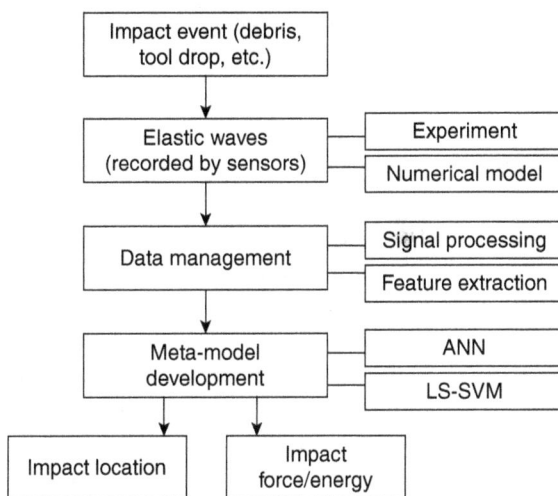

Figure 21. Flow chart for machine learning algorithms.

speaking, the radial base function (RBF) kernel is the most used method of non-linear function estimation.

The flow chart of both machine learning algorithms is shown in Fig. 21.

5. Optimization of Sensor Layout

The choice of the maintenance strategy for any structure will be decided by weighing the advantages of the proposed diagnostic system (e.g., reliability, probability of detection, monitoring of blind zones, fast interrogation) against its disadvantages (added weight and complexity, probability of false alarm and misdetection, sensor integrity under environmental and operational conditions). This is particularly important for condition-based maintenance (SHM systems) where the structural diagnosis and prognosis are carried out by utilizing the recorded sensor data. Therefore, optimal sensor number and location is a key parameter in the design of any SHM system. There are a number of published papers on optimal sensor positioning for damage detection.[37–41] A review of the proposed strategies for optimal sensor placement for impact identification is given in Ref. [42].

In most of the optimization algorithms, an objective function is either maximized or minimized. For impact detection and identification, example of an objective function to be maximized is the probability of detection and one to be minimized can be defined as the error function. The choice of the fitness function is very important as it has a direct influence on the optimal sensor location. It must be representative of the performance of the impact detection algorithm as well as the structural response. In Ref. [43] the proposed fitness function to be maximized is the Probability of Detection (PoD) function set as the probability that a sensor detects a strain higher than a given threshold. However, this is a very simplistic definition and it can be applicable for detecting an impact event, but not locating or characterizing the impact. Another disadvantage of some of the proposed methods is that often the response of the structure to an impact event is represented by simplified input models, which do not consider the geometrical or mechanical complexity of the structure.

For a large structure with a great possibility of sensor number and location, it is computationally expensive to carry out an exhaustive search to find the optimal sensor network. Therefore, genetic algorithm (GA) is often employed to reduce the computational cost of the problem.[44,45] Another proposed optimization strategy that reduced the computational cost of the problem is the GA-based optimization with the fitness function built on ANN and trilateration approach.[24]

In Ref. [46], a hybrid method for optimal sensor positioning based on modal parameters is presented for large-scale civil structures. However, the proposed algorithm is only valid for methodologies based on the modal response of the structure and its precision is directly related to the accuracy of the simulated dynamic response of the structure. In most of the available methodologies, the response of the structure is either simulated numerically or experimentally under controlled laboratory environment.

The sensor optimization, however, is not a deterministic problem and the effect of uncertainties associated to sensor data must be included in the optimization strategy. In Ref. [47], GA have been used with ANN by introducing uncertainty in terms of added noise to the sensor data. Another method to include the uncertainties in the sensor data is by adopting a Bayesian approach.[48] In addition, in the same work, the failure of one or more sensors as well as non-uniform probability of impact occurrence in the structure have been included in the optimization approach, which presents a more realistic scenario and increases the reliability of the decision-making system. The outline of this approach is presented in Fig. 22 and briefly described in this section and its application to a composite stiffened panel is reported in see Section 6.3.3 of chapter 7.

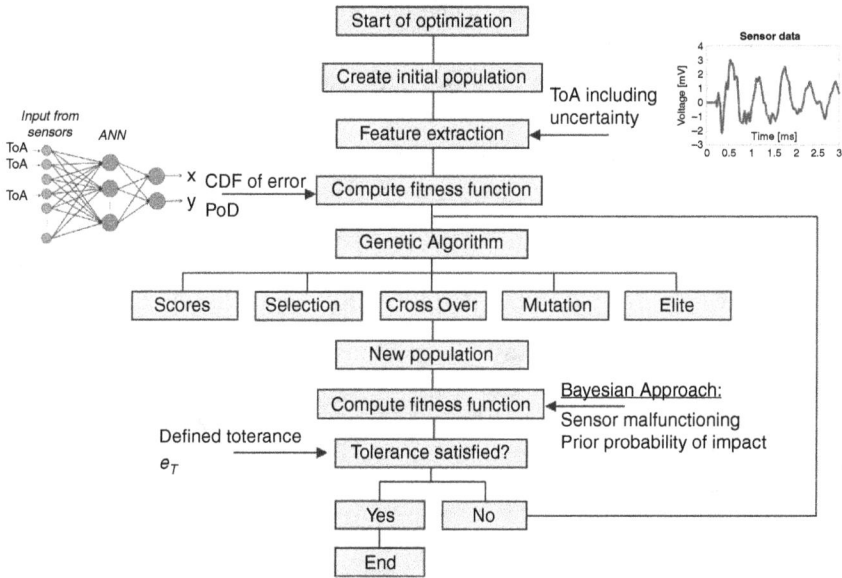

Figure 22. Flow chart of the Bayesian optimization procedure.

5.1. *Bayesian Optimization*

The Bayesian method proposed by Mallardo *et al.*[48] searches the optimum sensor location in order to maximize the PoD of an impact event in the presence of the probability of one or more sensors having failed and including uncertainty in the recorded data. Two main uncertainties were considered to influence the sensor data: (1) errors due to environmental and operational conditions such as instrumentation noise, bonding quality, temperature change and loading; (2) possibility of one or more sensor that may have failed during the operation. For example, let us consider the case where the input to the impact detection algorithm is in terms of time ToA of each recorded signal. The ANN will be trained and developed by a large library of impact scenarios (different locations, mass, velocity), which realistically will be obtained numerically. However, in real experiments, the ToA measures will contain uncertainties:

$$t_{inp,k} = t_{num,k} + \delta_1 + \delta_2, \tag{10}$$

where $t_{inp,k}$ and $t_{num,k}$ are the measured and computed ToA in the kth sensor respectively while δ_1 and δ_2 represent the variability and uncertainty in the data and are assumed as two independent Gaussian variables with zero mean and standard deviations σ_1 and σ_2. If the inputs $t_{inp,k}$ for $k = 1, \ldots, N$ (N being the total number of sensors) are independent, then the likelihood function for any k can be expressed as

$$p(t_{inp,k}|\boldsymbol{x}, \mu_k, \sigma^2) = \frac{1}{\sqrt{2\pi\sigma^2}}\exp\left[-\frac{1}{2\sigma^2}(t_{inp,k} - t_{num,k})^2\right], \quad (11)$$

where $\boldsymbol{x}$ gives the position of the impact, μ_k is the mean of the distribution and $\sigma^2 = \sigma_1^2 + \sigma_2^2$ is the total variance. Using the Bayes' theorem, the likelihood can be related to the posterior probability distribution function (PDF) and the prior distribution as

$$p(\boldsymbol{x}_i|t_{inp,k}) = \frac{p(t_{inp,k}|\boldsymbol{x}_i)p(\boldsymbol{x}_i)}{p(t_{inp,k})}. \quad (12)$$

The optimal sensor configuration is then obtained by minimizing the following problem:

$$\min_{s \in S} p\left[\bigcup_{i \in I}(\boldsymbol{x}_i|e_i > e_T)\right], \quad (13)$$

where $\boldsymbol{s}$ is a vector containing all the sensor readings, I is the library of impact events, each of which is located at $\boldsymbol{x}_i$ while $e_i = |\boldsymbol{x}_i^c - \boldsymbol{x}_i^p|$ is the error function relating to the predicted and the actual impact locations. The vector e_T represents a threshold value corresponding to the required performance of the SHM system in terms of the error function. The minimization given in Eq. (13) means that the optimal sensor configuration is searched in order to minimize the probability of not detecting and impact occurring at location $\boldsymbol{x}_i$. By applying the Bayes' theorem, the following minimization problem is obtained:

$$\min_{s \in S} \sum_{i \in I}[1 - p(e_i > e_T|\boldsymbol{x}_i)]p(\boldsymbol{x}_i) = \min_{s \in S} \sum_{i \in I}[1 - \mathrm{PoD}_i]p(\boldsymbol{x}_i). \quad (14)$$

As the impact events are independent, then the final probability can be calculated by the summation of the probabilities as denoted in Eq. (14). The second step, to increase the reliability of detection, is to include in the optimization algorithm the probability of one or more sensors malfunctioning at the instant of impact. If M_s represents the number of faulty sensors (defined by the user), the optimal sensor combination can be searched by minimizing the probability of an impact located at $\boldsymbol{x}_i$ not being detected within the acceptable error range $(e_i > e_T)$ as

$$\min_{s \in S} \sum_{i \in I} [1 - p(e_i > e_T | \boldsymbol{x}_i, M_s)] p(\boldsymbol{x}_i)$$

$$= \min_{s \in S} \sum_{i \in I} [1 - \text{PoD}_i] p(\boldsymbol{x}_i) p(M_s), \tag{15}$$

where the terms $p(\boldsymbol{x}_i)$ and $p(M_s)$ represent the prior probabilities of an impact occurring at $\boldsymbol{x}_i$ with M_s of sensors malfunctioning and are assumed to be independent probabilities. Therefore, by considering the probability of one or more sensors failing during the service life of the structure, the optimization process is improved. For more details on the implementation of the proposed optimization technique, see Ref. [48].

6. Application of Passive Sensing Algorithms

This section presents application of some of the introduced methods for passive sensing, resulting in impact location, categorization and contact force estimation using both experimental and numerical data on composite coupons and stiffened panels with surface mounted PZT sensors. Afterwards, the examples of the optimization technique resulting in the best sensor locations are also presented. The passive sensing techniques presented here are the ANN and LS-SVMs developed by Yue and Khodaei[23] and comparing the performance of each meta-model. The performance of the meta-model is defined by the error function which for impact location is defined as the mean detection radius for N number of impacts (test data):

$$r_{\text{mean}} = \frac{1}{N} \sum_{n=1}^{N} \sqrt{|x_n - X_n|^2 - |y_n - Y_n|^2}, \tag{16}$$

where (x_n, y_n) is the estimated impact location from the meta-model and (X_n, Y_n) is the actual impact location from test or numerical models. Error in the force prediction is defined as the percentage of difference between the maximum amplitude of the actual force F_n and the maximum amplitude of the predicted force f_n with respect to the actual force. For N testing data, the mean error is defined as

$$f_{\text{mean}} = \frac{1}{N} \sum_{n=1}^{N} \left| \frac{F_n - f_n}{F_n} \right| \times 100\%. \tag{17}$$

When an ANN is trained, it starts with a set of random initial weights. For this reason, every time a network is trained its performance can be slightly different. To obtain a robust representation of the network performance, each network is trained for 100 cycles by which the network behavior has already reached convergence,[45] and the best network is then saved. To report on the performance of a network, its probabilistic behavior is investigated by obtaining the PDF. It was shown that the distribution of the PDF function follows the lognormal type. Therefore, the mean and standard deviation of the longnormal PDF can best represent the ANN's response. Cumulative distribution function (CDF), which represents the PoD of the trained network against the mean error, can be obtained by integrating the PDF. To represent and compare the performance of each network, its means error at 90% PoD has been chosen as a good representation. This value can also be used in finding the optimal sensor number and position. For example, Fig. 23 shows the CDF for three different networks for estimating impact force magnitude.[9] It can be seen how the performance of the network at 90% PoD increases with the number of the sensors.

For the ANN results presented in this section, the total data set is divided into 50% training set, 25% validation and 25% as testing set.

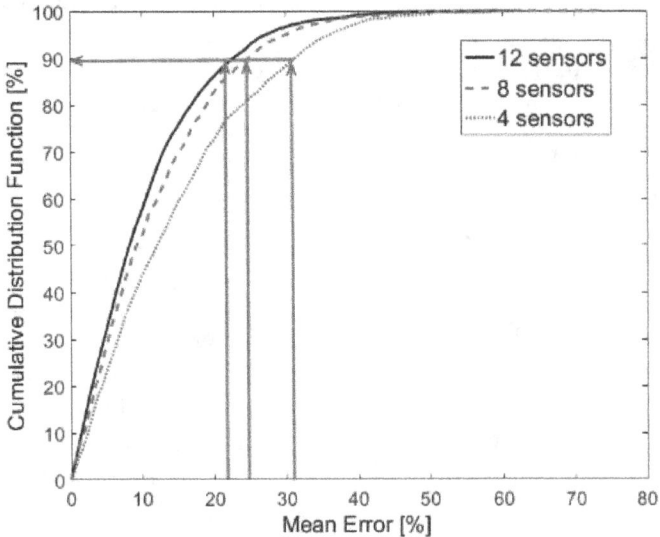

Figure 23. Probabilistic behavior of the error function for different sensor networks.

For LS-SVM,LS-SVM 75% of the data are used for training and validation and 25% for testing.

6.1. *Feature Extraction and Signal Processing*

The recorded sensor signals are large arrays of discrete time-variant data, which cannot be used as inputs to ANN and LS-SVM models directly. The features extracted for impact location is the ToA of the recorded signal, following a detailed case study.[25] To take into account uncertainties and noise in the data, Hilbert transform of the signal has been applied to find its envelope. The ToA of the signals is then defined as the time when the signal exceeds a certain threshold. This threshold depends on the sensitivity of the sensors, the minimum detectable strain as well as the level of the noise in the system (hardware) and the background. Considering signal noise level of 0.02 V and signal amplitude of about 15 V in,[23] the threshold for ToA was set to 0.2 V.

6.2. *Coupon Results*

A composite panel of size $300\,mm \times 225\,mm$ and layup of $[0/+45/-45/90]_{2s}$, with eight PZT sensors (PIC 225 from PI ceramics) was tested experimentally with 100 impact locations (see Ref. [23] for more details on the experimental set-up). Impact energy was kept low to avoid damage. A spherical impactor of 40 mm diameter has been used with three different masses: 2.7, 7.9 and 19.4 g each with 2.66 m/s velocity labelled as set 1, 2 and 3, respectively. Sensor readings have been recorded for each impact location and used to construct meta-models for impact detection and location. Signals were recorded by NI PXI1-1085 data logger. The recorded impact data from all eight transducers were then used to develop various ANNs and LSS-SVMs and to compare their performances. The ANNs developed for passive sensing are all multi-layer feedforward perceptron (MLP).

6.2.1. *Impact Localization Results*

Both ANN and LS-SVMs have been developed for locating impact on the composite coupon. The developed ANN consists of an input layer, one hidden layer with 20 neurons and an output layer with two neurons: x and y coordinates of impact. The ANN is trained for 100 cycles and the average performance is reported. The transfer function used in the training phase is "tansig" and the training algorithm is the conjugate gradient backpropagation with Fletcher–Reeves updates (CGF).

Since LS-SVM is only capable of a single output, two models are established for the (x, y) coordinate. The optimal kernel function for LS-SVM was selected from the detection accuracy of the LS-SVM models. Three different kernel functions were trained by a random set of data: linear, polynomial and RBF kernels. It was shown that RBF yields the most accurate estimation.[23] For the presented examples, due to the moderate size of the data set the cross-validation approach was selected.

For both meta-models, ToA is used as input. The first case study is related to regularization. When meta-models were developed with data from a single impactor mass, a better regularization was reached with LS-SVM (see the performance both in case 1 (2.7 g mass) and 2 (7.9 g mass) presented in Table 3).

The performances presented in Table 3 refer to the mean radial error as defined by Eq. (16). The following conclusions were made by comparing the performance of both meta-models:

- When the meta-models are trained with combination of set 1 and 2, the performance of both ANN and LS-SVM worsens in comparison to a single mass. However, for a small number of training set (case 3), the LS-SVM's error is much smaller than ANN.
- When the size of the training data set is doubled (case 4), ANN's performance improved noticeably.
- When the meta-model was developed using set 1 and tested with sets 2 and 3 (which have not been used in the training), LS-SVM showed a better performance.

It can be concluded that for small data set LS-SVM is superior to ANN. In addition, LS-SVM has a better generalization capability than ANN, which

Table 3.   Performance of the developed meta-models from experimental test.

Case	Training and validation set	Testing set	Performance of ANN (mm)	Performance of LS-SVM (mm)	No. of training and validation data
1	1	1	19.4	15.1	38
2	2	2	38.83	11.78	38
3	1, 2	1, 2	33.60	20.98	38
4	1, 2	1, 2	19.72	20.94	76
5	1	2, 3	52.67	36.16	38

makes it more suitable for applications where there is a great variation of the input data (i.e., various impact scenarios).

6.2.2. *Optimal Sensor Location Results*

Sensor topology (i.e., number and location) plays an important role in the performance of a developed meta-model. To find the best sensor distribution, an exhaustive search or an optimization analysis should be carried out.[19,42] In this section, an optimization strategy based on GA for optimal position of sensors for impact location is presented and applied to the composite coupon for validation.[49] For any optimization problem, a fitness function must be first defined, which will be either minimized or maximized during the process. In this case, the probabilistic error to locate an impact is introduced as the fitness function and will be minimized to find the optimal sensor topology. To evaluate the performance of each ANN, its response is simulated following the Monte Carlo approach with a certain number of samples. This results in a CDF distribution (see Section 6), which then gives a measure of the error corresponding to a defined PoD. During the optimization, for each possible sensor configuration, the probabilistic mean error related to a defined PoD will be measured and the sensor network with the lowest error will results in the optimal configuration. To validate that the proposed fitness function can best represent the performance of the network in an optimization, it was tested on the composite coupon, with eight sensors attached.[48] The goal was to find the best four-sensor network, out of eight, for locating impacts with high accuracy. As the total number of possible sensor combinations was limited to 70, for the sake of completeness, the fitness function for each sensor combination has been calculated and presented in Fig. 24. The horizontal axis presents all the 70 groups of sensors, while the vertical axis represents the proposed fitness function.

The PoD was computed based on the experimental sensor signals recorded during impacting the plate. To each recorded sensor data, 5% of noise has been added to introduce uncertainty. The sensor network with the minimum error is represented by a star and highlighted by a circle in Fig. 24. This network corresponds to the four corner transducers,[1,2,4,7] as expected, which are highlighted in Fig. 24. Once the fitness function is validated, the next step will be to apply the optimization strategy to a much larger problem such as a stiffened panel (see Section 6.3.3 of chapter 7).

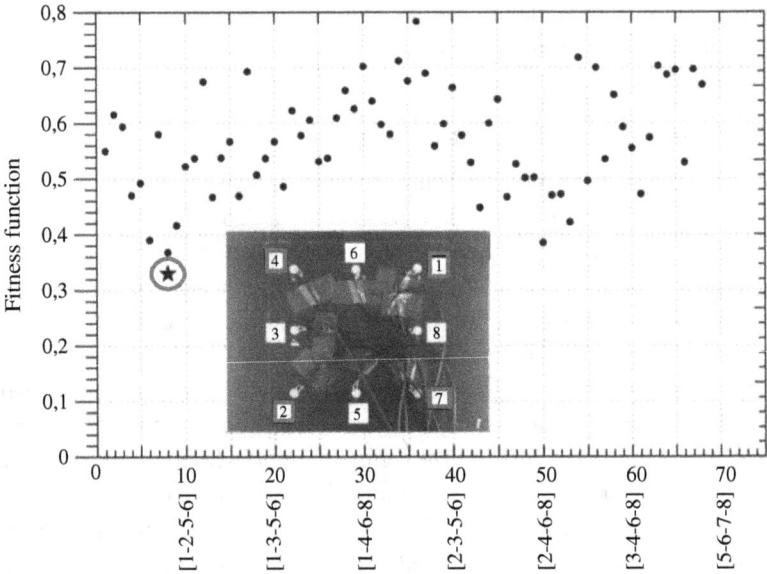

Figure 24.   Fitness function distribution for all possible four-sensor networks.

6.3. *Stiffened Panel*

To develop and train a neural network capable of identifying impacts on aircraft panels, a large number of sensor data from various impact scenarios, i.e., small mass (from outside) and large mass impacts (from inside), at various locations on the panel were simulated using a validate nonlinear numerical model (see Fig. 25). A total of 2,865 different impacts were simulated on a stiffened panel of 2,045 mm × 1,070 mm made of unidirectional woven carbon/epoxy composite plies. For more details on the impact scenarios, see Ref. [23]. There are 300 possible sensor locations in the panel. To compare the response of the ANNs and LS-SVMs, meta-models are developed for a four-sensor network, placed at the corners of the panel.[23]

6.3.1. *Impact Localization Results*

For both ANN and LS-SVM, the input is ToA of the four corner sensors and the output is the coordinates of the impact. The architecture of the ANN and LS-SVM is the same as the coupon test: feedforward MLP with one hidden layer of 20 neurons, transfer function: "tansig" and CGF training algorithm. For LS-SVM, fast-leave-one-out cross validation

Large mass impacts

Small mass impacts

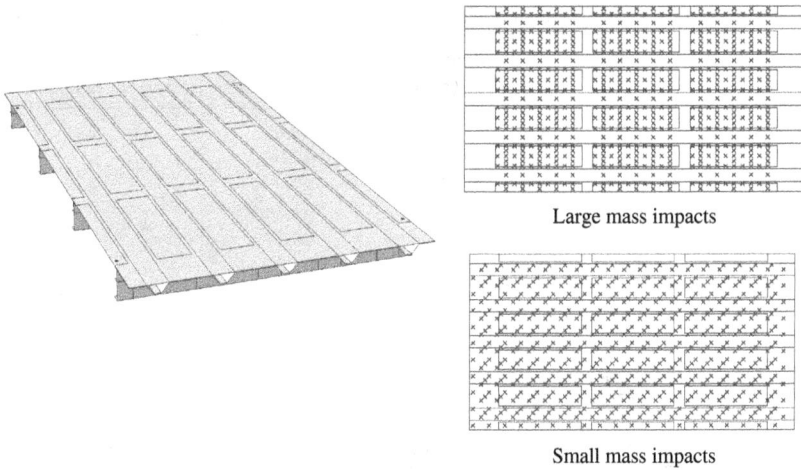

Figure 25.   Composite stiffened panel.

is used.[50] For this case study, since the size of the data set is much larger than the coupon test, the regularization and generalizations capabilities of both methods are further examined by models trained by large mass and small mass impact data separately and combined.

For both ANN and LS-SVM, the mean detection radius and the detectability radius corresponding to 90% PoD for the networks developed for large mass impacts are higher than those of small mass impacts. This is due to the fact that the response of the structure (and hence the sensor data) due to large mass impacts are much more predictable than for small impacts. Small mass impacts generate a much more random response in the structure and consequently the recorded sensor data, which means building a meta-model to predict this highly non-linear response is more challenging.

LS-SVM has a higher performance in comparison to ANN for both large mass and small mass networks, as presented in Table 4.

For the case of combined impacts, ANN and SVM have very similar performances. This indicates that when the size of the data set increases, ANN's regularization improves. To further investigate this, meta-models were developed and trained using the same amount of data (combined small and large mass impacts). The data presented in Table 5 confirms that LS-SVM has a better performance when trained with smaller data set in comparison to ANN. However, if the number of the training data is increased, their performance for impact detection is comparable. It also

Z. Sharif Khodaei & M. H. Ferri Aliabadi

Table 4. Performance of meta-models for impact location trained by different data sets.

Impact type	Meta-model	Mean radius (mm)	90% PoD radius (mm)	No. of training and validation data
Large mass	ANN	28.13	46.65	484
	LS-SVM	18.19	33.05	
Small mass	ANN	12.97	25.92	480
	LS-SVM	11.38	23.34	
Combined	ANN	14.26	26.33	942
	LS-SVM	14.35	27.86	

Table 5. Performance of meta-models for impact location — the size of data sets.

Impact type	Meta-model	Mean radius (mm)	90% PoD radius (mm)	No. of training and validation data
Combined	ANN	20.25	39.67	482
	LS-SVM	15.33	29.46	
Combined	ANN	14.26	26.33	942
	LS-SVM	14.35	27.86	

shows that by increasing the size of the data set by a factor of 2, the performance of LS-SVM does not improve.

The influence of the data set on the performance of ANN versus LS-SVM can be visualized in Fig. 26. The error of LS-SVM is one quarter of ANN when 5% training and validation data are used. As the data set is increased to 8%, the error of both models decreased dramatically. After small fluctuations, LS-SVM's error reaches a plateau at 15% of data, whereas ANN's error stabilizes with 33% of data.

The number and position of sensors greatly influence the performance of the ANNs. The main objective of testing different sensor patterns is to study the influence of the number and position of sensors on the performance of the trained ANNs, and therefore to provide a preliminary guidance on the optimization of the sensor network. The performance of two sensor networks, one with four sensors in the middle of the stiffened panel is compared to an eight-sensor network with sensors places at the edges of the panel.

The distribution of the detection radius over the full panel for the two different ANNs is presented in Fig. 27. It can be seen that impacts that occur outside the sensor network have much higher error in predicting their location. This effect can be seen as well for the impact locations

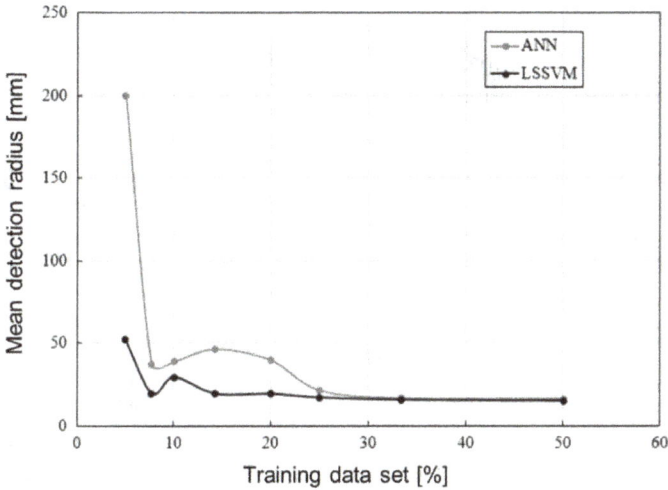

Figure 26.   Mean detection radius against training data set percentage.

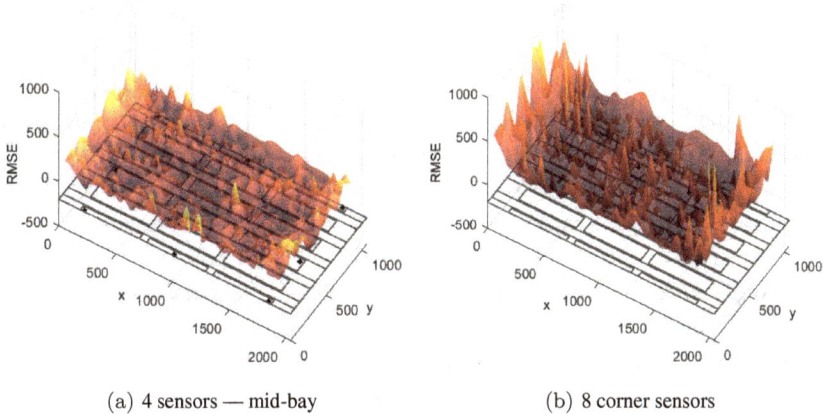

(a) 4 sensors — mid-bay

(b) 8 corner sensors

Figure 27.   Error distribution for impact location: Different sensor topologies.

close to the panel edge. This is related to the boundary reflected waves, which interferes with the sensor readings. Moreover, increasing the number of sensors results in error reduction. However, a more comprehensive optimization study is necessary to find the best sensor number and location for a given probability of detection, which will be covered next.

6.3.2. *Impact Force Detection Results*

Predicting the peak impact force is an important aspect of assessing the threats of impact on composite structures as it can be compared against the threshold for delamination. In this section, an example of identification of peak impact force using ANN and LS-SVM is presented for a composite stiffened panel.[23] Following the comprehensive studies on the optimal network architecture and input features for impact force detection,[9,51] the details of the developed ANN are the following:

- MLP feedforward network with scaled conjugate gradient (SCG) backpropagation training algorithm.
- Four signal features extracted as input to each sensor (four corner sensors): the maximum values of detailed and approximated coefficients of discrete wavelet transform (level 4 Daubechies wavelet), the maximum value of the sensor signal and its corresponding time; the output is the maximum contact force.
- Two hidden layer with 10 and 5 neurons.

LS-SVM models with fast leave-one-out validation are developed for peak impact force detection. Inputs and outputs for SVM are exactly the same as those for ANN. After investigating three different kernel functions, similar to the impact localization study, the RBF kernel resulted in the best performance and it is used for the development of the meta-models.

As it was pointed out in Section 2, the response of the structure to large mass and small mass impacts are significantly different, especially in terms of the contact force. Therefore, when the ANN was established for a combination of large mass and small mass impact data, the error in estimating the peak force amplitude is as high as 26% related to mean error and 61% error related to 90% PoD, for a large number of training data. The error was comparable in case of LS-SVMs as well (see Fig. 6). This error could be due to difficulties in reaching generalization, due to a large scatter in input and output data from both impact types. To investigate this further, separate networks have been trained for large mass and small mass impacts. The results presented in Table 6 show that the large error is attributed to small mass impacts.

The contact force for two different impacts carried out on a composite plate is presented in Fig. 28. Two experiments were carried out using a projectile with a head diameter of 25.4 mm, and mass and velocity of 6.15 kg and 1.76 m/s for the large mass impact using a drop tower; while

Table 6.   Performance of meta-models for max contact force.

Impact type	Meta-model	Mean error (%)	90% PoD (%)	No. of training and validation data
Large	ANN	10	23	484
mass	LS-SVM	8.7	18	
Small	ANN	35	60	480
mass	LS-SVM	21	39	
Combined	ANN	27	55	471
	LS-SVM	25	46	
Combined	ANN	26	61	942
	LS-SVM	22	41	

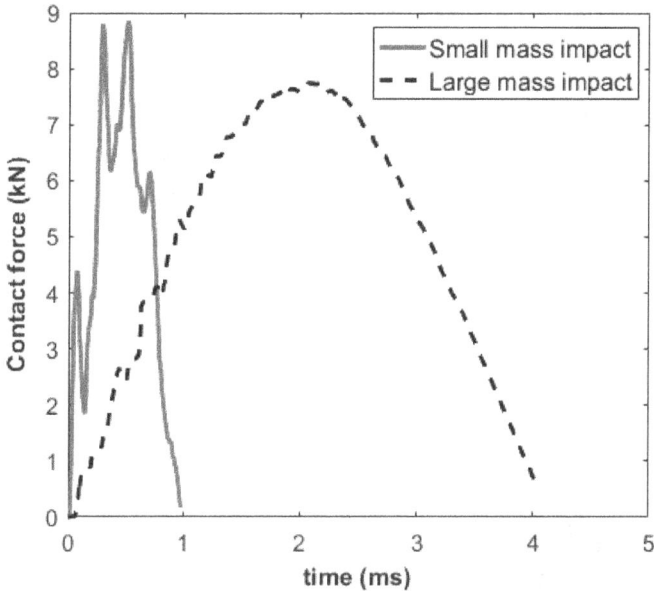

Figure 28.   Contact force versus time for small mass and large mass impacts.

the small mass impactor using a gas gun had 0.314 kg mass and 7.7 m/s velocity.[52] The contact force plots in Fig. 28 show the main differences in their behavior. It can be observed that small mass impacts have much smaller duration and cause a flexural wave controlled response in the plate. The contact force and structural response of large mass impacts can be represented with a few components of their frequency domain spectrum, but for small mass impacts many more spectral components are needed, which makes predicting the response of the structure very challenging, even using nonlinear models such and ANN and LS-SVM. The error of large mass impacts estimated with an ANN trained only for large mass impacts are less than half of small mass impacts. If two separate meta-models can be developed for each type of impact, the error attributed to the large mass impacts can be reduced. However, the passive sensing method must be capable of categorizing the impacts from the recorded sensor data.

The quasi-static and wave controlled responses of a plate subjected to large mass and small mass impacts suggest that spectral components of the strain data (structural response) also contain different energy distributions. For large mass impacts, the spectral sensor data contain large energies at lower frequencies. For the four-sensor network evaluated in this section, the spectral component of each sensor signal was determined for each impact and presented in a histogram plot in Fig. 29. The histogram also reveals a distinct threshold frequency between the two types of impacts (1,000 Hz). This criterion has been used for categorizing impacts into large mass and small mass impacts, based on the recorded sensor data, before the contact force is then estimated with the corresponding developed ANN or LS-SVM to further reduced the error in the contact force estimation.

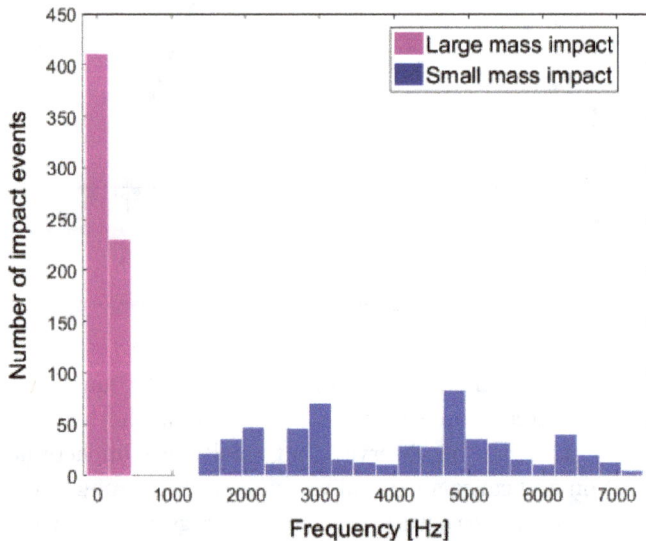

Figure 29. Dominant frequencies of all impacts: four corner sensors.

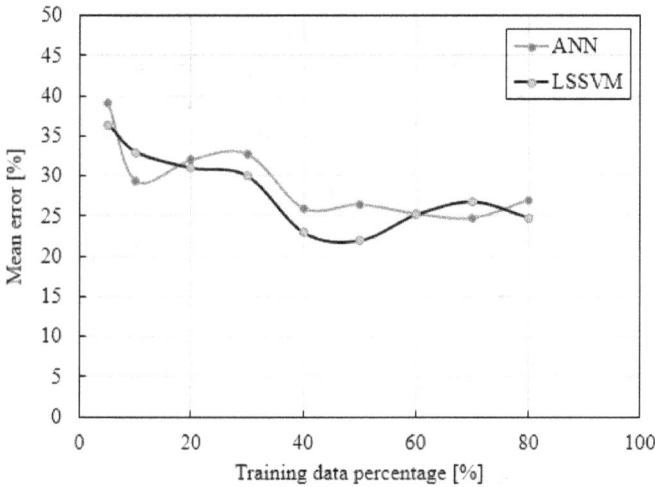

Figure 30.   Meta-model Performance against size of training set.

To compare the convergence of ANN and LS-SVM for contact force estimation, with regards to the size of the training data, the mean error (%) is plotted against the percentage of the training data with respect to the whole data set (see Fig. 30). Despite large fluctuations with small amount of training data, the performance of ANN stabilizes at around 45% of training data while LS-SVM's response reaches a local minimum around 45% of data but then increases. The performances of both models are similar and it is not clear which model performs better with a larger data set.

6.3.3. *Optimal Sensor Placement Results*

An optimization strategy based on GA is applied to the composite stiffened panel shown in Fig. 25 to determine the optimal number and location of sensors for impact localization. The fitness function to be minimized is the PoD function related to the probability distribution of the error for impact localization, validated in Section 2.2.

To reduce the computational size of the problem, a grid of possible sensor locations is assumed. This is more realistic representation of the problem because in real applications, there are several initial conditions defined such as the minimum distance between transducers, minimum edge distance and min/max number of transducers. In addition, under operational conditions, there will be uncertainties in the sensor data, possibility of sensor failures and non-uniform probability of impact

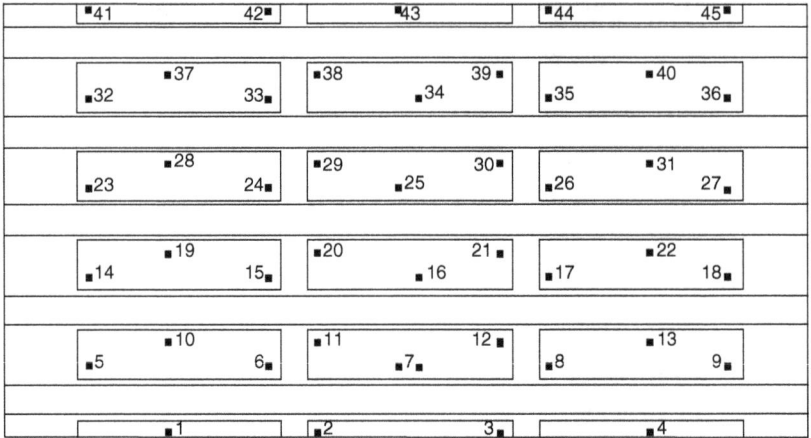

Figure 31.   Stiffened panel with all possible sensor locations.

occurrence[48] depending on different parts of the structure. Therefore, the Bayesian-based optimization introduced in Section 5.1 is carried out, which takes into account the uncertainty in the sensor data (environmental effect, bonding quality, instrumentation noise) and probability of one or more sensor failures.

In each bay, there are three possible sensor locations resulting in maximum of 45 possible locations as shown in Fig. 31.

The first test case was to find the best four sensor locations when the panel has a uniform probability of impact occurrence everywhere. The details of the GA can be found in Ref. [49] The defined threshold for the error in locating impacts is set to $e_T = 50$ mm. An initial population of 200 individuals is set. At each generation of the GA, a new population of 200 individuals is generated from the previous one by retrieving four elite individuals, mutating with 20% probability and cross-over of 80% of the parents. The optimal sensor combination with and without probability of sensor malfunctioning ($M_s = 1$ refers to probability of one sensor failure) is presented in Table 7. The results further strengthen the conclusion that the best performance is achieved when the sensors are places at the corners of the plate, maximizing the coverage area. To validate the choice of the optimal sensor network, its performance has been compared to similar sensor network, with similar coverage areas.

Table 7.   Comparison of the best four-sensor combination for the stiffened panel.

Sensor combination	Performance (%) $M_s = 0$	Performance (%) $M_s = 1$
[1 4 41 45] Optimal, uniform probability	83.1	74.1
[1 9 41 45]	82.8	73.8
[4 5 41 45]	82.8	73.8
[1 25 41 45] Optimal, non-uniform probability	84.0	—

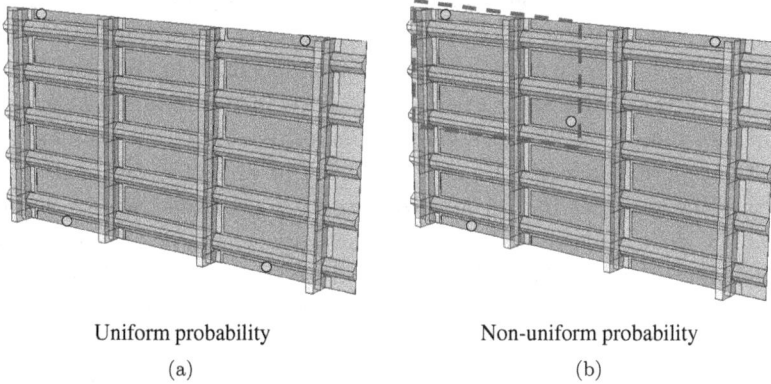

Uniform probability	Non-uniform probability
(a)	(b)

Figure 32.   Best four sensor network. (a) uniform probability of impact occurrence, (b) non-uniform probability of impact occurrence: higher probability in the highlighted region.

The results in Table 7 show that the performances of the sensor networks are very similar, but the optimal one is slightly higher. Furthermore, as expected, their performance deteriorates with the probability of having one failed sensor.

So far in all the examples presented, it has been assumed that the structure has uniform probability of impact occurrence. However, for real structures, this is not the case. For example, for an aircraft, there is 7% probability of impact occurrence on a fuselage section (without door) in comparison to 13% for wings and 31% for passenger door surroundings. To test this, the top left quarter of the stiffened panel has been given 80% probability of impact occurrence while the rest of the panel has 20% chance of being impacted. This area is highlighted with the dashed boxed in

Fig. 32(b). The optimal solution for the case with non-uniform probability resulted in removing one sensor from the corner of the plate and adding it to the area with high probability of impact. This increases the PoD of impact being detected in this area, which confirms the previous findings.

7. Conclusion

In this chapter, different SHM methodologies for passive sensing were introduced. Many of the proposed algorithms have been successfully applied to small, simple coupons in controlled laboratory environments. However, the focus of this chapter was on methodologies with the potential to be applicable to real complex structures under operational conditions. A special attention was given to machine learning algorithms (ANNs and LS-SVMs), which were established for impact localization and peak impact force prediction. The application of the developed meta-models on both composite coupons and large stiffened composite panels were demonstrated and their performances compared in terms of impact types and size of the training sets. For impact localization, LS-SVM was shown to be more suitable when the data set for training a network is limited. This reduces the computational cost of the problem significantly. By increasing the size of the training data set, ANN's performance was shown to match that of LS-SVM. It was also noticed that the error in localizing small mass impacts are much higher than large mass impact due to the nature of the response generated in the structure. Therefore, a categorization technique was proposed to classify impacts to small mass and large mass from the recorded sensor data. For the case of estimating the peak of the contact force, the error of the meta-models was rather high, due to the complexity of the problem. Similar to impact localization, the error for small mass impact was higher than large mass. In general, LS-SVM gave better results for single impact type and a better performance when trained with smaller data set.

The importance of developing a strategy for optimum sensor positioning was described. The Bayesian-based optimization algorithm was shown capable of determining the best selection of sensors in complex stiffened panels with both uniform and non-uniform probability of impact occurrence. It was shown how the performance of the network is changed based on different sensor configurations highlighting the importance of the optimal sensor placements and inclusion of probability of sensor failure in the optimization to increase the reliability of the HM system under real operational conditions.

References

1. Michaels, J. E. and T. E. Michaels, Guided wave signal processing and image fusion for in situ damage localization in plates. *Wave Motion* **44**(6) (2007) 482–492.

2. Michaels, J. E., A. J. Croxford, and P. D. Wilcox, Imaging algorithms for locating damage via in situ ultrasonic sensors. In *Sensors Applications Symposium, 2008. SAS 2008*, IEEE, 2008.

3. Giurgiutiu, V., *Structural Health Monitoring with Piezoelectric Wafer Active Sensors*. Academic Press, USA, 2007.

4. Flynn, E. B. *et al.*, Maximum-likelihood estimation of damage location in guided-wave structural health monitoring. In *Proceedings of the Royal Society of London A: Mathematical, Physical and Engineering Sciences*, The Royal Society, 2011.

5. Sharif Khodaei, Z. and M. H. Aliabadi, Assessment of delay-and-sum algorithms for damage detection in aluminium and composite plates. *Smart Materials and Structures* **23**(7) (2014) 075007.

6. Olsson, R., Mass criterion for wave controlled impact response of composite plates. *Composites Part A: Applied Science and Manufacturing* **31**(8) (2000) 879–887.

7. Faivre, V. and E. Morteau, *Damage Tolerant Composite Fuselage Sizing* in *FAST — Flight Airworthiness Support Technology (48)*, August 2011, Airbus. pp. 10–16.

8. Olsson, R., Analytical prediction of large mass impact damage in composite laminates. *Composites Part A: Applied Science and Manufacturing* **32**(9) (2001) 1207–1215.

9. Ghajari, M. *et al.*, Identification of impact force for smart composite stiffened panels. *Smart Materials and Structures* **22**(8) (2013) 085014.

10. Morse, L., Z. S. Khodaei, and M. H. Aliabadi, Reliability based impact localization in composite panels using Bayesian updating and the Kalman filter. *Mechanical Systems and Signal Processing*, **99** (2018) 107–128. Elsevier, https://doi.org/10.1016/j.ymssp.2017.05.047

11. Giubergia, R., *Structural Health Monitoring: Environmental and Impact Variability on Passive Sensing* in *Aeronautics* Imperial College London, 2016.

12. Tobias, A., Acoustic-emission source location in two dimensions by an array of three sensors. *Non-Destructive Testing* **9**(1) (1976) 9–12.

13. Coverley, P. and W. Staszewski, Impact damage location in composite structures using optimized sensor triangulation procedure. *Smart Materials and Structures* **12**(5) (2003) 795.

14. Ciampa, F., M. Meo, and E. Barbieri, Impact localization in composite structures of arbitrary cross section. *Structural Health Monitoring* **11**(6) (2012) 643–655 doi: 10.1177/1475921712451951.

15. Seydel, R. and F.-K. Chang, Impact identification of stiffened composite panels: I. System development. *Smart Materials and Structures* **10**(2) (2001) 354.

16. Seydel, R. and F.-K. Chang, Impact identification of stiffened composite panels: II. Implementation studies. *Smart Materials and Structures* **10**(2) (2001) 370.

17. Inoue, H., J. J. Harrigan, and S. R. Reid, Review of inverse analysis for indirect measurement of impact force. *Applied Mechanics Reviews* **54**(6) (2001) 503–524.

18. Ghajari, M., Z. S. Khodaei, and M. H. Aliabadi, Impact detection using artificial neural networks. In *Key Engineering Materials.* Vol. 488–489, pp. 767–770, (2012), Trans Tech Publications, Switzerland.

19. Staszewski, W. *et al.*, Fail-safe sensor distributions for impact detection in composite materials. *Smart Materials and Structures* **9**(3) (2000) 298.

20. LeClerc, J. *et al.*, Impact detection in an aircraft composite panel — A neural-network approach. *Journal of Sound and Vibration* **299**(3) (2007) 672–682.

21. Vapnik, V., *The Natural of Statistical Theory.* Springer-Verlag, New York, 1995.

22. Fu, H., *et al.*, Fast detection of impact location using kernel extreme learning machine. *Neural Computing and Applications* **27**(1) (2016) 121–130.

23. Yue, N. and Z. S. Khodaei, Assessment of impact detection techniques for aeronautical application: ANN vs. LSSVM. *Journal of Multiscale Modelling* **7**(4) (2016) 1640005.

24. De Stefano, M. *et al.*, Optimum sensor placement for impact location using trilateration. *Strain* **51**(2) (2015) 89–100.

25. Sharif Khodaei, Z., M. Ghajari, and M. H. Aliabadi, Determination of impact location on composite stiffened panels. *Smart Materials and Structures* **21**(10) (2012) 105026.

26. Schumacher, T., D. Straub, and C. Higgins, Toward a probabilistic acoustic emission source location algorithm: A Bayesian approach. *Journal of Sound and Vibration* **331**(19) (2012) 4233–4245.

27. Yan, G. and J. Tang, A Bayesian approach for localization of acoustic emission source in plate-like structures. *Mathematical Problems in Engineering,* **2015** (2015), Article ID 247839, 14 pages, http://dx.doi.org/10.1155/2015/247839

28. Zárate, B. A. *et al.*, Structural health monitoring of liquid-filled tanks: A Bayesian approach for location of acoustic emission sources. *Smart Materials and Structures* **24**(1) (2014) 015017.

29. Niri, E. D. and S. Salamone, A probabilistic framework for acoustic emission source localization in plate-like structures. *Smart Materials and Structures* **21**(3) (2012) 035009.

30. Moon, Y.-S. *et al.*, Identification of multiple impacts on a plate using the time-frequency analysis and the kalman filter. *Journal of Intelligent Material Systems and Structures* **22**(12) (2011) 1283–1291.

31. Yan, G., H. Sun, and O. Büyüköztürk, Impact load identification for composite structures using Bayesian regularization and unscented Kalman filter. *Structural Control and Health Monitoring* **24**(5) e1910 (2017) doi: 10.1002/stc.191.

32. Suykens, J. A., Support vector machines: A nonlinear modelling and control perspective. *European Journal of Control* **7**(2) (2001) 311–327.
33. Wong, P. *et al.*, Engine idle-speed system modelling and control optimization using artificial intelligence. *Proceedings of the Institution of Mechanical Engineers, Part D: Journal of Automobile Engineering* **224**(1) (2010) 55–72.
34. Xie, J. Kernel optimization of LS-SVM based on damage detection for smart structures. In *2nd IEEE International Conference on Computer Science and Information Technology, 2009, ICCSIT 2009*, IEEE, 2009.
35. Vong, C.-M., P.-K. Wong, and Y.-P. Li, Prediction of automotive engine power and torque using least squares support vector machines and Bayesian inference. *Engineering Applications of Artificial Intelligence* **19**(3) (2006) 277–287.
36. Van Gestel, T. *et al.*, Financial time series prediction using least squares support vector machines within the evidence framework. *IEEE Transactions on Neural Networks* **12**(4) (2001) 809–821.
37. Guo, H. *et al.*, Optimal placement of sensors for structural health monitoring using improved genetic algorithms. *Smart Materials and Structures* **13**(3) (2004) 528.
38. Flynn, E. B. and M. D. Todd, A Bayesian approach to optimal sensor placement for structural health monitoring with application to active sensing. *Mechanical Systems and Signal Processing* **24**(4) (2010) 891–903.
39. Gao, H. and J. L. Rose, Sensor placement optimization in structural health monitoring using genetic and evolutionary algorithms. In *Smart Structures and Materials*. International Society for Optics and Photonics, 2006.
40. Guratzsch, R. F., *Sensor Placement Optimization Under Uncertainty for Structural Health Monitoring Systems of Hot Aerospace Structures* Citeseer, 2007.
41. Thiene, M., Z. S. Khodaei, and M. H. Aliabadi, Optimal sensor placement for maximum area coverage (MAC) for damage localization in composite structures. *Smart Materials and Structures* **25**(9) (2016) 095037.
42. Mallardo, V. *et al.*, Optimal sensor placement for structural, damage and impact identification: A review. *SDHM: Structural Durability & Health Monitoring* **9**(4) (2013) 287–323.
43. Markmiller, J. F. and F.-K. Chang, Sensor network optimization for a passive sensing impact detection technique. *Structural Health Monitoring* **9**(1) (2010) 25–39.
44. Worden, K. and W. Staszewski, Impact location and quantification on a composite panel using neural networks and a genetic algorithm. *Strain* **36**(2) (2000) 61–68.
45. Mallardo, V., M. H. Aliabadi, and Z. S. Khodaei, Optimal sensor positioning for impact localization in smart composite panels. *Journal of Intelligent Material Systems and Structures* (2012) doi: 1045389X12464280.
46. Yi, T. H., H. N. Li, and M. Gu, Optimal sensor placement for structural health monitoring based on multiple optimization strategies. *The Structural Design of Tall and Special Buildings* **20**(7) (2011) 881–900.

47. De Stefano, M. *et al.*, On sensor placement for impact location: optimization under the effect of uncertainties. In *Proceedings of Fifth European Workshop Structural Health Monitoring*, 2010.

48. Mallardo, V., Z. S. Khodaei, and F. M. H. Aliabadi, A Bayesian approach for sensor optimisation in impact identification. *Materials* **9**(11) (2016) 946.

49. Mallardo, V., M. H. Aliabadi, and Z. S. Khodaei, Optimal sensor positioning for impact localization in smart composite panels. *Journal of Intelligent Material Systems and Structures* (2012) doi: 1045389X12464280.

50. Ying, Z. and K. C. Keong, Fast leave-one-out evaluation and improvement on inference for LS-SVMs. In *Proceedings of the 17th International Conference on Pattern Recognition, 2004. ICPR 2004*. IEEE, 2004.

51. Yue, N., Impact detection in sensorized composite panels. In *Aeronautics*. 2015, Imperial College London, 2015. p. 45.

52. Pierson, M. O. and R. Vaziri, Analytical solution for low-velocity impact response of composite plates. *AIAA Journal* **34**(8) (1996) 1633–1640.

53. Mills, N. J. and A. Gilchrist, Finite-element analysis of bicycle helmet oblique impacts. *International Journal of Impact Engineering* **35**(9) (2008) 1087–1101.

54. Mills, N. J. *et al.*, FEA of oblique impact tests on a motorcycle helmet. *International Journal of Impact Engineering* **36**(7) (2009) 913–925.

55. Johnson, K. L., *Contact Mechanics*. Cambridge University Press, UK, 1985.

56. Yang, C. *et al.*, Some aspects of numerical simulation for Lamb wave propagation in composite laminates. *Composite Structures* **75**(1–4) (2006) 267–275.

57. Su, Z. and L. Ye, *Identification of Damage Using Lamb Waves: From Fundamentals to Applications*. Springer-Verlag, Berlin, 2009, x, p. 346.

Appendix A: Impact Simulation

In Section 2, the importance of variability of impact scenarios when generating a transfer function or a meta-model for impact detection and characterization was highlighted. In general, it is not realistic to assume that the dynamic response of the structure under an impact event will be obtained experimentally for a large number of impact scenarios. Therefore, the importance of valid numerical techniques to simulate an impact event on a composite structure and generated ultrasonic wave is stressed in the development of SHM methodologies. In this chapter, a validated numerical technique is presented where the key parameters to ensure an accurate model are described. ABAQUS FE software has been employed to model a composite stiffened panel after validating the model at coupon level.

Aircraft panels are usually composed of a skin and several stringers and stiffeners, which increase the global stiffness of the structure as well as its critical buckling load. Given the small thickness of these components, they can be adequately discretized with shell elements. In addition, due to the small thickness of the skin the response of the structure will be non-linear due to large deflection. This is an important factor in modeling impact as the nonlinear response should be captured. A suitable shell element in ABAQUS is the reduced integration general-purpose shell element (S4R). For a composite panel, each lamina can be represented with an integration point defined through the thickness of a shell element. Orientation and material properties of the lamina are assigned to this point, thus eliminating the need for using one element for each lamina. The impactor, which is usually assumed to be much stiffer than the panel, can be meshed with discrete-rigid-elements.

Contact Definition

The definition of contact at the impactor–panel interface directly influences the prediction of the impact force history. FE implemented contact algorithms usually generate master and slave surfaces from contact pairs and predict their closure and separation. If a node of the slave surface penetrates into an element of the master surface, the contact algorithm enforces constraints to reduce the penetration or to keep it at zero. In the penalty contact method, a spring is inserted between the slave node and the penetration point on the master element, whose stiffness is based on the stiffness of the underlying elements. When the aim is to maintain the penetration at zero, the contact is named *hard*. Conversely, the *softened* contact refers to contacts for which a pressure–overclosure relationship is to be satisfied at the slave nodes.

Using hard contact between a rigid impactor surface and shell elements, spikes with very high magnitudes appeared in the contact force.[53,54] Alternatively, a softened contact was employed with a pressure–overclosure relation derived using the following Hertzian relation[55]:

$$F = k_{\mathrm{H}} d^{3/2}, \qquad (\mathrm{A}.1)$$

where F is the total contact force, d is indentation and k_H is the contact stiffness. The impactor is modeled with rigid elements while conventional shell elements have been used for meshing the skin. The overclosure of the nodes of the impactor, which are slave with respect to the panel, represent the indentation. The details of calculating the contact stiffness and the indentation can be found in Ref. [25] where both the indentation and the contact force are validated against experimental results for a composite plate.

Sensor Modeling

Once a valid model of impact simulation is available, the next step is to simulate the sensor response. Once an impact event occurs, elastic waves are generated in the structure which, after propagation, is recorded by surface mounted PZT sensors. The sensors are capable of sensing in-plane strain through the generation of electrical charge Q (direct piezoelectric effect):

$$Q_i = d_{ijk}\sigma_{jk} + p_{ij}K_k, \tag{A.2}$$

where d is the piezoelectric coupling coefficients, p is the permittivity, K is the electric field and σ is the induced stress. There is a linear relationship between the electric charges accumulated on both surfaces of a PZT sensor and the sum of the in-plane strain components measured at the sensor–plate interface.[56] Since the sensor is relatively small, variation of the strain components over the sensor area is negligible and therefore the average strain of sensor elements can represent the sensor strain. PZT transducers are usually modeled with elastic solid elements.[57] Given the small size of these transducers (not more than 10 mm in diameter), often their required element size reduces the stable time step of the explicit solution and consequently increases the simulation time drastically. Since a large number of impact simulations should be carried out to provide data for ANNs, it is worth employing an alternative method for strain measurement without modeling the sensor.

A suitable alternative could be removing the sensor and measuring the strain at integration points of skin's elements located in a square surrounding the sensor.[25] To compensate for the reduced stiffness at this location, one integration point is added to each shell element, whose mechanical properties are the same as those of the sensor elements but its assigned thickness is $\pi t_{\text{sensor}}/4$. The validation of the sensor model was carried out by impact simulation on a composite plate and recording the sensor data with modeling the PZT sensors with sensor element and the proposed integration point model. The result of in-plane strain and resulting voltage presented in Fig. A.1 validates the proposed sensor model against the 3D model of the sensors, which results in noticeable reduction in computational time.

Once the contact algorithm and the sensor model have been established, the impact model on a composite plate with two surface-mounted sensors was simulated and compared with experimental results. The features that are used in passive sensing are amplitude of the contact force and the ToA

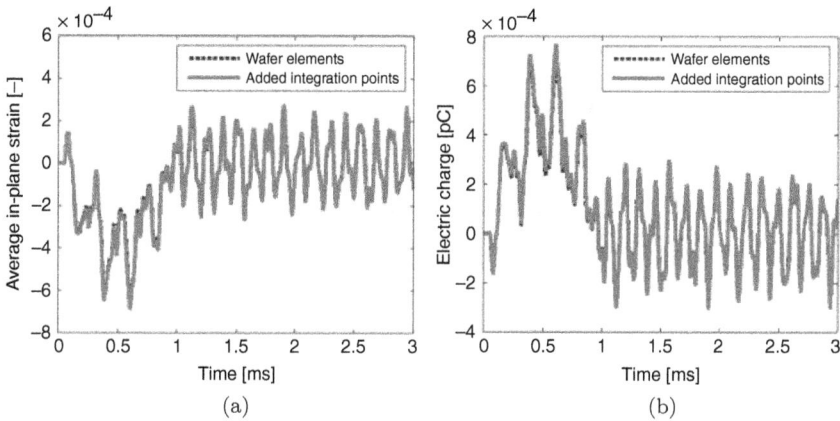

Figure A.1. (a) Average in-plane strains recorded by a wafer and added integration points and (b) the corresponding electric charges.

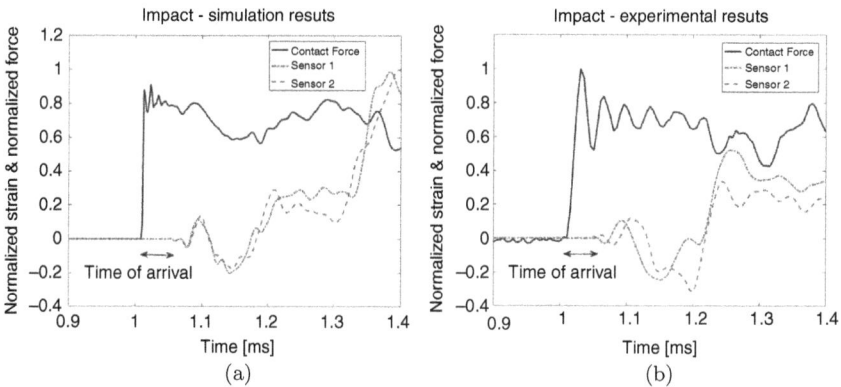

Figure A.2. Validation of impact simulation on a sensorized composite plate. (a) Numerical results and (b) experimental results.

of the recorded sensor data. Therefore, the two features have been validated for the developed numerical model (see Fig. A.2).

Even though in some cases there are moderate discrepancies between FE and experimental results, it can be concluded that the accuracy of the predictions of the FE model is adequate.

Index

A

absorbing boundary, 83, 85
absorbing boundary condition (ABC), 52, 82, 88
absorbing layers with increasing damping (ALID), 47, 83–85, 88
acceleration, 55
accellent, 161
Acellent SMART Layer, 149
acoustic emission, 199
acoustic emission (AE) source, 228–230
acousto-ultrasonics (AU), 212
actuation frequency, 170, 172
adhesive bond, 128–131, 146
admittance, 128–134
analytical solution, 52
anisotropic materials, 7
antisymmetric Lamb modes, 48
antisymmetric mode, 5–7, 49, 71
antisymmetric SH modes, 48
aperture, 177
aperture angle, 173
artificial neural network (ANN), 28, 215, 234–237, 240–241, 243–250, 252, 254, 258
attenuation, 7, 22, 29, 50, 79, 81, 165
auxiliary points, 55

B

backpropagation, 236, 252
bandpass filters, 181

barely visible impact damages (BVIDs), 2, 22, 34, 35, 215
baseline, 182
baseline signal stretch (BSS), 15
baseline-free, 12, 25–26, 34, 40
Bayes, 242
Bayesian, 16, 215, 237–238, 240, 256, 258
Bayesian optimization, 241
Bays risk, 28
beam forming, 153, 157, 168, 174
bond assessment, 146
bonded, 206
bonded joints, 126–127
boundary conditions, 153, 157
boundary element method (BEM), 96
boundary reflections, 82

C

C-scan, 207–208, 210
carbon fiber reinforced composite (CFRP), 8
Cauchy's integral, 110
CFRP isogrid structure, 204
CGF, 246, 248
charge simulation technique (CST), 96
Christoffel's equation, 100
Christoffel's operator, 110
Christoffel's solution, 100
circular frequency, 5, 169
comparative vacuum monitoring (CVM), 212

Computational and Experimental Methods in Structures

(Continued from page ii)